Doing Science

CORWIN
PRESS

The Corwin Press logo—a raven striding across an open book—represents the happy union of courage and learning. We are a professional-level publisher of books and journals for K-12 educators, and we are committed to creating and providing resources that embody these qualities. Corwin's motto is "Success for All Learners."

Doing Science

Innovative Curriculum for the Life Sciences

Neal A. Glasgow

CORWIN PRESS, INC.
A Sage Publications Company
Thousand Oaks, California

For information address:

Corwin Press, Inc.
A Sage Publications Company
2455 Teller Road
Thousand Oaks, California 91320
E-mail: order@corwin.sagepub.com

SAGE Publications Ltd.
6 Bonhill Street
London EC2A 4PU
United Kingdom

SAGE Publications India Pvt. Ltd.
M-32 Market
Greater Kailash I
New Delhi 110 048 India

Printed in the United States of America

Library of Congress Cataloging-in-Publication Data

Glasgow, Neal A.
    Doing science : innovative curriculum for the life sciences / author, Neal A. Glasgow.
        p.   cm.
    Includes bibliographical references and index.
    ISBN 0-8039-6477-3 (pbk. : acid-free paper). — ISBN 0-8039-6476-5 (cloth : acid-free paper)
    1. Life sciences—Study and teaching (Secondary).   I. Title.
QH315.G54   1996
574'.071'2—dc20   96-25243

This book is printed on acid-free paper.

96 97 98 99 10 9 8 7 6 5 4 3 2 1

Production Editor: Gillian Dickens
Designer/Typesetter: Christina M. Hill

Contents

Preface

How many of us have actually done science? We read about science. We watch television programs about people that do science. We hear about the new knowledge and discoveries that are the outcome of science activities, but most of us have not had the opportunity to participate in the process of investigation and discovery that makes science so exciting.

Today, there are trips called "ecotours," which are vacations that offer people trips to remote rain forests, archeological dig sites, or ocean locations to work with scientists as technicians. Most of the trip involves and engages participants in gathering data for a research scientist and his or her study. The researchers interact with participants, teach techniques, and share their knowledge with people. It all makes a very effective way to experience how science works, to enjoy oneself, and to learn something at the same time. People pay money to try to "touch" the scientific process. What a motivating and interesting learning experience.

What makes this experience different from a secondary science classroom? There is a growing division between the way science is taught as a series of facts or history and the way science is being conducted in graduate science programs, working science laboratories, and investigative field study sites worldwide. Many classroom activities today support only preordained outcomes and the teaching of existing knowledge. The term *hands on* or *mind on* and related jargon do not guarantee that students have experienced science as a dynamic process of scientific inquiry—the scientific method. To experience the power of the scientific method, individuals need to have the opportunity to create new knowledge exploring original questions and engaging in the creation of investigative pathways that lead to answers. To most, the textbook is science. The science in the textbook and related "canned" laboratory experiences are to be learned, mastered, and regurgitated come test time. Engaging in the true process of "doing" science is inherently more interesting than just learning about science. So how do educators move beyond the textbook, typically used as curriculum, to having students engage in a more authentic and motivating science experience?

What you will find here is the working philosophy, methods, and examples behind the idea of problem-based instruction to move past textbook-guided life science curriculum. It does not do away with the textbook, but puts it to work applying its contents in the form of information that students need to apply to more authentic science applications or activities: doing science in science class. This is a kind of problem-based learning that blends the learning of the traditional knowledge within:

- An investigative or exploratory framework engaging students in the act of really doing science with true unknowns
- The requirements of creating the necessary investigative or exploratory frameworks
- Collecting and analyzing the generated data to find answers to original questions

The idea of acquiring knowledge, technique, and process within loosely structured learning problem scenarios or original investigative research projects creates multiple learning pathways with the flexibility to meet the needs of a wide variety of learners within the same class.

This pedagogical style and curriculum has its roots in the scientific method and a problem-solving framework. It uses and exploits the natural curiosity and creative talents of students to get them out of their seats as active participants, not passive recipients of instruction. This style of learning and teaching appeals to any science educator looking to put the act of doing science back into the secondary classroom.

This book was not designed as an activity or laboratory book. However, there are numerous examples of activities or ideas to engage students that can be implemented with few modifications. What it was designed to do is to open up opportunities to expand an instructor's teaching "toolbox."

The book synthesizes one science program's 5-year curricular development experience into a form that addresses and reflects all the concerns one faces when involved in curricular innovation and change. Professionals involved in curricular innovation and change will find not only philosophy but practical information that will help them deal with and better understand the issues they confront. They include

- Overview and critique of secondary science education
- Definitive identification of many current instructional or curricular strategies
- The creation, design, and development of problem and activity scenarios
- Student management methods within activities
- Curricular accountability within problem-based activities; matching activities to framework guides
- Assessment and evaluation methods; examples and suggestions on how to do it within problem-based activities
- A large number of working applications and examples of activities; many offer examples of how a particular activity might really work in a class
- Communicative strategies for managing curricular change for students, teachers, and parents

There is no one definitive answer or ideal curriculum style for all classrooms. Only by knowing all the options can an educator be free to choose a style that complements their circumstances and personal skills and meets their students' needs. It is hoped that the information contained here empowers educators with the insight, options, and opportunities that allow them to construct the most effective classroom atmosphere and environment possible.

Introduction to Science Teaching Philosophy

Looking at an Old Paradigm
With New Eyes

When I was young I would visit a local reservoir, called Lake None. It appeared very large at the time. We built rafts and pushed through the cattails looking for birds, fish, and amphibians. We collected tadpoles, frogs, toads, crayfish, and minnows, took them home, set up mini-habitats and observed nature in the informal ways many young people do. I happened upon a rattlesnake striking, killing, and slowly swallowing a rabbit. Playing tug-of-war with trapdoor spiders as they fought to keep the lids on their burrows created the question about why they had their burrows here and there and not somewhere else. Going to the beach and finding chitons would draw associations with ancient trilobites, a connection I could not explain. Then came junior high, and science became a book, worksheets, and laboratory experiences with right and wrong answers, memorizing vocabulary, taking notes, and listening to the teacher. This pattern did not change much in high school. The "investigations" were not explorations at all. They were directions to "right" answers that we should follow. Everything supported the facts, vocabulary and general scientific history that was presented in the "science encyclopedia," the textbook. I did not realize until after I had taught science for a number of years, that I was doing science as a child and only studying about science as a student, and teaching science history as a teacher. Being a good student, I began my teaching career by perpetuating the patterns that I experienced as a student, concerned about coverage and the "hands-on" phenomena. Students learned how to be good science students by following the directions and the preordained paths and outcomes that I designed for them. The popular teachers developed the skills of showmanship, taking center stage and trying to drum up student enthusiasm for the day's factoids and "cookbook" labs. The personal intrinsic values of curiosity and wonder seemed to disappear. Science no longer contained the motivating factors of self-discovery and the unknown. Many times grades became the greatest motivating force. What happened to the natural curiosity and imagination in science? Where did it go?

This is a book that explores curricular design and development in science education, specifically education strategies in the life sciences. Unlike most science activity books, this book's contents strike a balance between curricular ideas and philosophy and the concrete steps, guidelines, and real activities that reflect the outcome of curricular change within one successful science program.

For some, it offers a new look at an old paradigm: classroom science teaching. In Chapter 1, "Teaching and Learning Today: The Basic Styles and Options," various curricular philosophies and teaching styles are discussed. Eventually, readers are led to a curricular style, pedagogy, or delivery system that traces its roots to the basic premise or foundations of science: asking questions, investigating and researching, working through experimental design, and problem solving. It is commonly called problem-based learning but also has been called process, investigative, or inquiry-based learning.

Chapter 2, "The Basic Elements or Characteristics of the Problem-Based Curriculum," helps define this curricular paradigm and goes on to suggest elements within any problem, theme, or topic that may make it appropriate for curricular development. Chapters 3 through 7 explore a range of curricular activities that engages students, student groups, and full classes in the problem-based approach to science teaching and learning. These problems, themes, topics, and curricular units range from structuring authentic science investigations, both on and off campus, to interesting, open-ended simulations.

All activities offer unlimited opportunities to scale the rigor and complexity to a variety of learner levels and resources within the same class. In some cases, students may desire more from a science program than a class can provide. Many science students entering competitions, such as the Westinghouse Competition, Duracell Competition, or Science Fair require expertise and equipment that the school cannot provide. For those students and their teachers, Chapter 5 covers the task of expanding the classroom by enlisting the local science community. Local businesses, agencies, colleges, and universities engaged in doing science offer students tremendous opportunities. The recruitment of area mentors as well as project management and assessment are all covered.

Finally, acknowledging real change is very difficult. Chapters 8 through 11 support changes in curricular and pedagogical strategies by exploring assessment evaluation philosophy; curricular accountability; parents and staff; and classroom strategies for individual, group, or cooperative study. These chapters respond to many of the questions that typically are expressed by those involved—students, other teachers, parents, and school administration—during the implementation of new curricular programs.

How Do They Relate? The Evolution of Modern Science and Today's Science Classroom

Science is in an interesting position in today's society, and one can find similar or parallel evolving public perceptions in the world of art. Walking through any modern or contemporary section of an art gallery or museum, you can overhear clichéd criticisms and comments in response to a communication breakdown between the artist and the public. Modern art does this to some people. Sometime around the turn of the 19th to 20th century, cutting-edge artists began to explore and communicate not just the visual world, apparent to most of us, but a world of intellect and cognition. The objects of their study, interest, and exploration could not be seen and understood as easily as in the past. Art became intellectually interesting and relevant, not just an aesthetic or exercise in visual realism. Nonrealistic and abstract art became the artists' response to their discovery and exploration of the cognitive psychological or sociological realm. Art media, the materials of art expression, expanded beyond paint and pencil to include just about everything and anything. Once the visual "obvious" was exhausted as subject matter, other areas, less visible and more intellectual, needed to be considered for investigation and visual interpretation by the art world.

Most people can look at art based on an artist's interpretation of visually realistic subject matter and critique its success or failure from an aesthetic sense. They may not be able to explain the work verbally, however. The intent of the artist is easy to understand for them. In today's contemporary art world, art requires the viewing public to be much more proactive in the intellectual exercise of becoming "art literate." Literacy in any realm develops throughout life with no end point that can be attained. It is a continuum developed throughout life. The number of steps to literacy in many areas has grown in number, and the steps have become steeper. The breakdown of communication between the artist and the public began.

Without some mutual thread or common experiences to facilitate connections to specific artwork, opportunities for potential understanding become further removed from the viewing public. Time is important. As people walk by any art exhibit, how much energy are they willing to commit to becoming engaged in the pursuit of understanding? Without time to reflect on possible intellectual connections to the art, communication breaks down. Whether the artist has an understanding of this phenomenon may not be important. Is it the artist's responsibility to explain the intent of the artwork, or should the work stand on its own without interpretation? These are philosophical arguments about the responsibility of the art world to aid in creating an art-literate public.

Science is now, more than ever, in the same position. For most of the general public, contemporary science is an intellectual exercise. Most people cannot connect to what is going on in today's science laboratories. There are fewer mutually shared reference points, common experiences, or threads of understanding connecting the public to the subjects of modern research. People are not able to see, hear, feel, or touch the focus of most research or the techniques used. During a recent visit to the life sciences building at the State University of New York at Stony Brook, I walked up five floors. Each floor contained a biological specialty, from Genetics to Ecology and Evolution. The vast majority of the observed laboratory research included molecular biology techniques using data collection strategies very different from laboratory techniques used in most secondary classrooms. Even in Ecology and Evolution, research was being done on the ecology and conservation of specific genes and their expression in different sea urchin species. In English, students can write in the same styles and use the same techniques as the playwrights, authors, and poets they study. Math teachers can bring the most complex problems to their students to work through. Math problems that engineers and others routinely deal with, and the electronic calculators and computers to solve them, can be available to students. This is not the case with science. Life science teachers have a hard time duplicating or practicing authentic science and teaching concepts and principles within current context—and not just as a history or body of knowledge. When the science classroom cannot connect to the true reality of the way science is now practiced, it is no wonder the general science literacy gap is expanding.

In some cases, the science that the public sees has moved into the realm of ethical or moral philosophical thought, argued from emotion, beliefs, or fears because of the lack of concrete connections to current scientific understanding. Science remains mysterious. The steps to science literacy, like art, have become steeper, and there are more of them. This pathway to science literacy is not static but very dynamic, and few are willing to put in the work needed to gain understanding.

Does the Science Classroom Connect?

In the past, public school classrooms used to be the place to develop the connections needed to understand and experience science. The study of art has become an educational sideline in most public school settings. Science has not. Science literacy is still a goal and expectation of public school experiences. Ideally, science literacy is defined as an integrated, interdisciplinary understanding of the function of science in the scope and scale of societal activity. This concept is in addition to the general understanding of specific science content and knowledge.

For a variety of reasons, much of what goes on in today's science can no longer be experienced in the public school classroom. The relationship and role of science within our society become elusive. The expense of the required equipment and potentially dangerous materials, or the lack of teacher training or understanding, make up the greatest reasons for the growing gap between what is really happening in science and what is being taught and experienced in the science classroom.

What Is Science Literacy, and From Where Does Science Curriculum Come?

Defining the word *literacy* means looking at a range of reading and writing abilities. The range of reading and writing competencies can be measured, seen, and heard. Defining science literacy is more difficult. It seems every science education group concerned with education is developing standards, frameworks, goals, and objectives to guide teachers in selecting or creating meaningful educational practice. They all contain the current jargon—major principles, concepts, collaborative learning, major themes, process skills, hands-on, minds-on, and so on. This is better than the content or factoid outlines of the past. Their structures sound good and, for a while, motivating, until you try to build real lessons and activities using them. The guides get put on the shelf along with the guides from 5 years ago. So, if the curriculum does not come from these guides and a teacher's interpretation of them, from where does it come?

It comes from textbooks. Are the textbooks current? Do textbooks reflect the curriculum guidelines from all these sources? Do they reflect the latest educational methodologies, and do all school classrooms have them? The answers to these questions may be disturbing, with estimates of 75% of biology classroom time and 90% of the homework involving the use of textbooks (Blystone & Barnard, 1988). Textbooks become the curriculum in most classrooms. An informal, anecdotal review of 17 high school biology textbooks (Jablon, 1992) found little change in the structure of biology textbooks over the past 30 years. The reviewer went on to say that current science educational research does not support the type of science teaching methodology implicit in the structure and content of the textbooks (e.g., breadth vs. depth or cookbook labs).

The guiding light of good intention becomes a blinding glare of confusing philosophical rhetoric from so many directions with little connection to the reality of the classroom. Textbooks describe science as a compendium of facts and processes, and they have become "biological encyclopedias." If literacy is defining how scientists do science or understanding how science is a way of knowing or how it is applied in the real world, most science classrooms and curricula miss the point of what literacy is. "Investigations" are not designed to teach students to ask good questions, design experiments, create hypotheses, control variables, gather and use data, assess reliability, and so on. They are conducted to point out preordained phenomena and lead students, if they follow the recipe, to the "right" answer. Most often, science is taught as a body of knowledge, techniques, and maybe concepts—a history, so to speak—to be interpreted by the teacher for the student. Much of the time, relevancy and context apply to the individual class with few connections to the world outside the classroom. Students rarely engage in any authentic science process.

The complexity and amount of scientific knowledge grow quickly. The tools that science and scientists use to "see" and work with what they study are, many times, not available to or appropriate for a classroom. Becoming literate has now become a dynamic interactive process as modern science continues to generate, modify, and replace knowledge while explaining the structure and function of living and physical matter, energy, and patterns. A compendium of current facts is not enough. There is a slow movement toward hanging scientific knowledge in the major concepts of science. Focusing on a framework of concepts may increase retention and understanding of detailed knowledge but does not ensure a relevant context of applied understanding.

Today, hidden information within the natural world is explored on a scale, both large and small, not even dreamed about years ago. Within all this mix is the way science is connected to the human condition and its function within society. The science curriculum has not been able to keep up with the demands that the modern definition of science literacy has put on it. How do the students fit into the picture? Structured science educational experiences become selective and homogeneous for interest, motivational, and perceived ability levels. *Rigorous* and *challenging*, defined in many classroom settings, mean simply more information or knowledge and a faster pace. If this type of class is deemed a valid pathway to literacy, does it appeal to all students? Do *rigorous* and *challenging* mean they will become more literate more quickly? Rigorous and challenging careers in science and in general are defined very differently in the real world. They involve looking and working with complex problems and developing investigations, not recalling content for tests. Maybe *challenging* and *rigorous* should mean the same thing in both settings.

Certain rigorous classes sometimes become social status symbols for teachers and students. What is on the transcript can become more important to students than the experiences within the class. The perception of rigor or challenge and the success of self-motivated students do not validate a curriculum. Less engaged students receive different experiences for a number of reasons. Literacy outcomes are defined differently for them. Yet they will vote and have to consider the same issues as adults as will the more engaged students. Lack of engagement does not always reflect a lack of potential, and these students may become very motivated later in life when they do not have to play the "school game."

Long-held perceptions of common educational jargon may limit identification of curriculum issues, which also limits reform and perpetuates educational myths. The terms *academic, vocational,* or *technical* each bring to mind a stereotypical view of the type of activities that might go on in a science classroom setting using these terms as descriptors. Yet the same pedagogy—applying knowledge and information to carefully chosen authentic problems—might be just as valid a strategy in auto shop as in advanced placement biology. Do "academic" students learn that much differently from "vocationally" or "technically" oriented students? Many times, the differences between academic students and nonacademic students are not innate ability levels but only motivational, interest, or social expectation levels. When we speak in educational jargon, associated stereotypical paradigms may hinder creativity and innovation in deciding on teaching and learning methods. I always liked the reality of art and shop classes because assessment and evaluation were based on doing something with what you learned, not just taking a test on it. An "A" meant you mastered what you needed to know and did something with it. A product was produced that demonstrated it. Does an "A" mean the same thing in all classes? Consider these issues carefully when planning for a changing curriculum or adapting new instructional strategies.

I have been fortunate to find adult associations with people who have made science their careers. I define myself as a science voyeur. I enjoy knowing what they do and how they do it. In some capacity or another, these people are engaged in self-discovery, exploration, and investigation as adults. They create investigations to answer questions for themselves and others in their chosen discipline. They do what I did as a child. The complexity and scale have changed, and the tools are more sophisticated. There are no right or wrong answers, only greater or lesser reliability within experimental design, data, and conclusions. What they do is very different from what happens in most science classrooms today.

I recently accompanied a graduate student to her laboratory. That day she was doing micro-injections of sea urchin mRNA into frog oocytes to try to stimulate the oocytes to produce a protein used as a sperm receptor. I had heard the idea that genes are not specific but more universally shared and found throughout all living things. Now I was seeing evidence for this phenomenon in these tiny eggs. Her laboratory was my Lake None. Collecting and working with DNA, RNA, and proteins and understanding their behavior takes great skill, knowledge, and understanding of complex techniques. She needs these tools. Somehow, regardless of her classroom science experience, the spark of curiosity and self-discovery was still there. She found some personal relevance to keep her going. Nature and natural phenomenon still hold wonder, spawn questions, and create the unknowns for her to investigate and try to understand and explain.

Science, as practiced in the real world, is a search for answers to questions about and within the natural world. Many children practice this informally and with little knowledge but lots of curiosity. Scientist and graduate students in the sciences practice this more formally in laboratories or other research settings. They too retain their sense of curiosity and have opportunities to satisfy their curiosity through original investigation and research. What happens to students in biology classrooms between pre-school and the graduate level is a mixed bag very dependent on a variety of factors. Too often, the experiences may not reflect the building and retaining of relevant scientific literacy useful in any context other than that specific classroom.

Common sense tells us that assimilating all the current biological knowledge and related vocabulary in textbooks or cookbook laboratory experiences would not define a biologically literate person. Understanding the major concepts helps, but those concepts have to be able to be applied to real issues and contribute to real-world scientific literacy. Textbooks do not portray an accurate view of the processes in science. What do students learn by just memorizing the textbook? In most cases, they do not know how the knowledge was obtained or about the people who obtained it. They could not tell you if the knowledge would be useful in their lives or whether it would be relevant today or in the future. They would not learn it in the investigative, process, or research context in which it was discovered. In most cases, they would not know how it applies to current scientific inquiry, how it influences social or global issues, or how it connects to other sciences and disciplines. They would never get to experience and practice science themselves. If scientific literacy means understanding how scientists do science or the characteristics of scientific knowledge, they would never be considered literate. The arts are far more accessible than the processes of the sciences, and the arts today are misunderstood.

Where does that leave science in the public school setting? First, textbooks and canned, cookbook laboratory experiences should not be considered curriculum leading toward the type of scientific literacy required in today's world. The power of science as a way of knowing may never be experienced or understood. The culture of teaching influences the student's lifelong perception of science, and this, in turn, influences the culture at large. This influence may determine the direction of society and the future of science. Curiosity—naturally found in children and a necessary characteristic of graduate researchers and scientists—should not be driven out of public school science and replaced with a content-, knowledge-, or information-based experience. A curious student is a much more enjoyable experience for the teacher than a bright student driven by grades. It is the spirit of curiosity that should be the single most important motivating factor in the drive for information. Science classrooms need to foster it and use it.

Biological literacy is understanding the nature of scientific inquiry and understanding a small number of pervasive scientific and biologic principles or concepts. This includes applying them in appropriate ways when reading a newspaper, listening to the radio, watching television, engaging in discussion, and seeking and interpreting scientific knowledge and information while making personal and societal decisions. Exposing students to large amounts of vocabulary and content does not ensure retention or foster interest and motivation in any but a few students.

The implications of this definition of literacy for the classroom are the paring down of content and content-driven activities and providing students with ample opportunities to build a deeper understanding of concepts connected to the real world. These opportunities can be built to use curiosity as the engine or the power required to acquire overall literacy. Curiosity can be channeled into the scientific inquiry process. A creative, investigative, open-ended curriculum, wrapped around pathways of self-discovery beyond the textbook, can be designed and developed to facilitate the acquisition of knowledge, concepts, and principles while experiencing the process of doing science in a relevant and authentic context.

We have discussed many reasons and justifications that support a reexamination of the current way in which science is experienced by many students. Educational inertia has made it difficult to replace current classroom practices with those more appropriate to the facilitation of lifelong, ongoing scientific literacy over a dynamic timeline. This inertia has been created by our own experiences as students and the behavior and modeling of our peers. Our idea of what a teacher and student are expected to be and do is engraved from our experiences and reinforced by what we have seen in the classrooms of our peers.

A problem-based investigative curriculum offers the potential for design and development of a range of learning experiences that supports dynamic scientific literacy. In the spirit of the problem-based approach, the ideas here are discussed and presented as triggers and clues for readers to adapt for their own purposes. It may not be something you can take into your classroom on Monday morning, but, in the long run, it will provide a new curriculum delivery system more relevant to today's educational demands.

The State of the Science Classroom

Science, as experienced in the secondary science classroom, has lost touch with the way science is practiced and used in our society. Textbooks and related activities, laboratory experiences (usually purchased from catalogs), and activities using out-of-date techniques contribute to science educational experiences that are very much out of context with the reality of the way science is done. Preparing new science teachers for classroom management and training them to design and develop a state-of-the-art curriculum is difficult. Classroom management and curricular design do not have to be mutually exclusive tasks.

Science has a backbone of questioning, investigation, and research in which everything else—all knowledge, processes, and techniques—connects and is used. Unfortunately, this idea is missing in much of the curriculum in secondary science classrooms. Art has managed to split the history of art, what we know about art, from the process of learning to produce artwork. This has not happened in the science classroom. Simply put, authentic science experiences are rarely used or practiced in classrooms. There are creative individual teachers out there doing some really great things, but their innovative approaches and programs are not adapted into mainstream reform or change. Without their energies, or when they leave their schools, their programs die.

So how do we better connect the reality of today's science with curricular design and development? There are no formal universal protocols for this to happen despite the many well-intentioned science educational philosophers who have good ideas. In actuality, their ideas usually do not evolve into concrete curricular methodologies and instructional practices ready to be implemented. Nationally, science curriculum application becomes haphazard and random as teachers fill school days with activities. There are many reasons for this, and they are discussed in later chapters. Curriculum validity is rarely investigated in individual schools or classrooms.

Self-analysis, analytical reflection, innovation, change, and reform are not mainstream staff activities in most schools but can be part of an individual teacher's ongoing career evolution. These activities can be maintained and nourished by self-directed and motivated individual teachers. Educational inertia, created by the instructional or institutional bureaucracy and the provincial roles of students and teachers, slowly rolls over all but the most resilient. The majority of science educational activities experienced by teachers and students are due to passive acceptance of this inertia. Students' experiences become a disjointed collection of haphazard activities with few connections to their roles in the larger science schema.

The following narrative is written for those teachers who are searching for ways to resist inertia and who think innovation, changes, and reforms are relevant and in context to their personal goals of providing their students with the most useful, effective, relevant science education possible. Above all, it is hoped that the information will be intellectually challenging and will contribute to clarity when thinking and making decisions about your own classroom programs and science education in general. The activities and ideas presented here are examples of an educational approach that offers students a variety of open-ended educational pathways where curiosity, creativity, and using their minds will still count. Authentic questioning, investigation, and research balance, use, and enhance the acquisition of science literacy. Students and the scientific realm unite in a more genuine context. An integrated and interdisciplinary literacy is the goal.

This book is an attempt to identify and define current practices and present alternatives that may meet the needs of a wider range of students in facilitating scientific and biological literacy. It offers examples of concrete activities and ideas that can be used to meet expectations presented in the better scientific literacy standards and frameworks while retaining interest, motivation, relevancy, self-discovery, and curiosity in correct contemporary scientific context. Take out of this book, either from ideas or examples, what you think could work for you. Modify it, use it, and make it your own.

Defining the Jargon: The Glossary

Educational jargon means different things in different settings. An example of this is the term *portfolio*. Over the past few years, I have seen many seminars with the term portfolio in their titles. Each presenter had created his or her own unique species of portfolio definition. After trying to describe the theme of this book to others outside of education, I realized that there may be a communication problem. Like the word portfolio, the term *problem-based learning* may not describe much of anything to others out of the jargon loop. Look in the back of any chapter in most textbooks. There are lots of problems there, yet they would rarely be of value in the problem-based context of this book. Because of this phenomenon, I have created my own definition for some of the jargon that may apply only in the context of this book. This is not all of the jargon by any means, but it is a start. Some words are defined and discussed later in the text.

Problem-based learning is a phrase that describes gaining and acquiring knowledge and information and learning technique and processes while working toward the solution to a problem, investigation, or the production of a product. In our framework, the problem activity itself creates and defines the learning and teaching paradigm. Knowledge, information, techniques, skills, and processes gain instant relevant context and meaning within the problem's framework. The problems and problem-solving processes become the curricular and instructional methodologies.

A *mentor* is a person, not technically a teacher, interested in providing the same things that teachers provide for students. Mentors usually make their living in some profession other than education.

Projects, in the context of this book, are interesting and motivating problem-based, educational pathways or experiences that are designed and developed by mentors or teachers to impart students with the knowledge and skills necessary to be successful in their careers.

Self-directed is a term that describes students taking a greater, more active role in their own education. It is also a goal and objective of problem-based pedagogy.

A *passive learner* describes a traditional learning and teaching style. The teacher is active and directs most of the learning processes. The student's role is defined by the teacher.

Active learners energetically strive to take a greater responsibility for their own learning. They take a more dynamic role in deciding how and what they need to know, what they should be able to do, and how they are going to do it. Their roles extend further into educational self-management, and self-motivation becomes a greater force behind learning.

Open-ended refers to educational experiences that have multiple educational opportunities and potential outcomes within the same project. A project will also have aspects that appeal to the many ability levels and interests within the same classroom. The process of learning is the focus, not the result.

Outcomes are project or educational goals and objectives as defined by students, mentors, and teachers. These may be very different because each project participant, including the teacher and mentor, have different outcomes in mind within the same project or problem-based activity paradigm. Outcomes may be work products, or they may be more conceptually based. Outcomes, in one context, become evidence of mastery of any given educational activity.

Frustrations include the "hoops" we have to go through to get to our outcomes. In problem-based learning, frustrations give us the opportunity to have students experience, with our guidance and help, what it is really like to try to get real work done.

Real world is a phrase that describes our effort to create greater student connections between the skills and knowledge being taught in the classroom and the skills and knowledge required in post-high school experiences.

Content, techniques, and *processes* include the more traditional stuff that we find in most textbooks, canned activities, and contrived, less authentic (as related to professional skills) school activities. In biology, content is the matter dealt with in the subject of biology or knowledge of the facts or history of biology. Techniques are the methods and tools that a biologist would use in gathering data or creating an experimental design. Processes refer to the thought processes involved in doing science, sometimes referred to as the scientific method. I do not view the basics—reading, writing, and math—as contrived, but they are usually taught out of context. The applications of these basic skills and the school work to which we have traditionally applied them may be less relevant and therefore may be seen by students as not being directly important to them.

About the Author

Neal A. Glasgow co-founded and teaches in a small, specialized, self-contained school-within-a-school emphasizing science, math, and technology in northern California. He has taught and been involved for 12 years in instruction program design at the middle school, junior high, and high school levels. The smaller specialized school is located within a larger restructuring high school, and he has been involved in development of instructional programs and curricular design for both schools. The school is widely known and recognized for its innovative instructional programs, especially its award-winning community mentor program. The school has received recognition as a runner-up for the State of California School Board Association's Golden Bell Award and received Sonoma State University's and Sonoma County Department of Education's Jack London Award for innovation and excellence in education. Its community mentor program has received many inquiries and visitors over the years and was the focus of a 1993 television program called *Classrooms of the Future*.

Glasgow has spoken at and provided seminars locally, regionally, and nationally on many aspects of educational program design. He is the author of two other books on learning and teaching strategies and curricular design: *Taking the Classroom Into the Community: A*

Guidebook and *New Curriculum for New Times: A Guide to Student-Centered, Problem-Based Learning*. His main goal is the creation of curricular activities that engage students in the most motivating, interesting, and valid curriculum possible.

1

Teaching and Learning Today
The Basic Styles and Options

Breaking down teaching and learning styles helps outline the options that teachers are offered. This chapter may be common knowledge to many of you, yet it is still worth reviewing while setting the stage for further consideration. The chapter's purpose is to summarize and define the basic educational models (Barrows, 1985; Barrows & Tamblyn, 1980; Kaufman, 1985) used in science classrooms and to establish a common background to which to refer. It is acknowledged that in actual practice, models can become more complex or can be combined or modified in any number of ways.

Two categories of teaching-learning can be defined here. The first categorization is based on the person responsible for making the decisions of what learners are to learn and how they are to learn it.

◪ Is it the teacher (teacher-centered) or the student (student-centered)?

The second category is based on how the body of knowledge and skills is organized.

◪ Does it center on subject areas (subject-based) or defined problems or problem areas (problem-based)?

Curricula can be teacher-centered/subject-based, student-centered/ subject-based, teacher-centered/problem-based, or student-centered/ problem-based. It also can be a combination of all or a few categories. The following describes briefly these educational systems or modalities.

Teacher-Centered Learning

In teacher-centered learning, the teacher is solely responsible for selecting what scientific information and skills the student is expected to learn, how they are to be learned, and the sequence and pace at which they are to be delivered. A student's science experience varies between teachers, classrooms, and schools. There are district guidelines and state guidelines, but teacher-centered educational experiences in high schools are not generally standardized. Therefore, teachers use science curricula that they know or feel comfortable teaching; thus, content coverage, related activities, and delivery style vary dramatically.

This is a well-known model that most of us have been exposed to since kindergarten. The teacher's role is a very traditional one in that in this method, the teacher's function is to dispense and interpret scientific knowledge and information via lecture, assigned readings, demonstrations, and selected activities. The teacher selects the resources, activities, and curricular delivery style. Multiple learning style approaches or opportunities may not be offered or given consideration. The teacher also sets the standards for assessment, evaluation, and the demonstration of mastery. The characteristic that identifies a teacher-centered curriculum is that a student is less directly responsible for his or her own education. Textbooks become the knowledge to be interpreted by the teacher. The students become accustomed to being passive rather than active recipients or participants.

Advantages

- ◪ The teacher can be certain that the student is exposed to all the knowledge and concepts that are appropriate for the targeted curricular unit.

It is easy for knowledgeable teachers to synthesize difficult subjects or topics into easily digested capsules, making this method the most efficient for dispensing content knowledge. This usually involves some kind of interpretive use of a textbook, to which the student can always refer if needed. It saves the student and the teacher the agony, frustration, and time that would be needed to structure learning and to work through the subject on his or her own.

- This method is universally recognized by students, teachers, parents, and administrators.

Success as a teacher in this format depends on the teacher's knowledge and his or her style for dispensing it. This style can be expressed and demonstrated in class organization, personal insight, humor in lectures, selected learning resources, and so on.

Disadvantages

- Students are not homogeneous in background, knowledge, or experience, nor are they homogeneous in learning abilities in different areas or in their pace and style of learning.

Each may have different career aspirations or levels of interest and motivation. In teacher-centered learning, the teacher imposes what he or she assumes all students should know with little regard to the heterogeneous needs of the class, curricular content, pace, or learning style.

- The student is generally a passive recipient of this model and does not "learn to learn."

The student's task is to learn what is offered and regurgitate on demand in a style of the teacher's choosing. If the teacher relies on a textbook, it may not be of appropriate reading level for all students. The content of the textbook may be outdated or otherwise inappropriate. The student's reward, usually external in nature and motivation, is based on grades.

- A false sense of security may be obtained by teachers, students, and parents.

They need to believe that the curriculum is valid, and once information is dispensed and a cognitive framework provided, students will incorporate the information. Without a more authentic context and relevancy, students may not recognize where and when it could or should be used and apply it effectively. An individual teacher's content background may not be current or may be based on textbooks that may not be relevant.

 ◩ This model assumes that the information is the most current, correct, and useful, and that the material is in a retainable format.

No one can predict which parts of the information that students have learned eventually will become obsolete or incorrect, nor what students will forget or retain. The abilities needed to find and evaluate new scientific information are not fostered.

 ◩ If the educational program is based on lecture-type learning, it is important to recognize that it cannot be delivered at the convenience of the learner, nor can it be given at a level, pace, and priority important for individual learners in the class.

Student-Centered Learning

In this method, the student learns to determine what he or she needs to know to find success within the class structure. Although the teacher may have considerable responsibility in facilitating investigative and discovery activities, it is expected that the student will take full responsibility for his or her own learning, with necessary experience and guided practice, until he or she gains full independence. The emphasis is on active student acquisition of information and skills suitable to ability, level of experience, and educational needs. The student decides the best individual manner of learning, resources necessary, and the pace and structure of acquisition within the activity. This is usually done in collaboration with, and with the facilitation of, the teacher. This may include choosing the style of the demonstration of mastery.

In both the student-centered and teacher-centered methods, the teacher may prepare what he or she feels are appropriate learning objectives, learning resources, and evaluation materials, and a choice of pathways that reflects his or her particular experience and knowledge. In the teacher-centered approach, these materials prescribe what a student is to learn. In the student-centered approach, these materials serve as guides and resources to be used and adapted as a student feels appropriate for taking responsibility for his or her education. A term paper, with the subject choice coming from the teacher, becomes a teacher-directed pathway. This is a conventional example of a combination of curricular approaches. With few requirements other than a subject, it could become a student-centered activity as the student is free to respond to the assignment in a manner of his or her own choosing. It is an open-ended project. Other than the subject, the rigor, depth of exploration, sources of data, authenticity and validity of support information and conclusions become the choice of the individual student.

Many of the projects in Chapter 4 are examples of the combination of curricular approaches. The *Island Project*, in which the students are required to answer orally some specific questions for guidance at the end of each section, is a great example of this. Students are free to learn the answers to these questions in any way they want. They are also free to illustrate their mastery of the questions in any manner. With no right answers, students are free to work at any level of complexity or rigor they desire. The freedom that these assignments offer does create insecurity in some students, and they usually ask for guidance. This is much different from "right answer" recipe assignments that have only one way to view the lessons or problems.

As teachers create appropriately relevant projects and problems, they provide choices for student exploration and investigation. These experiences put knowledge and skills in a more authentic context because students determine what they need to know and master in the process of finding solutions to problems, participating in projects, and meeting educational expectations and objectives. Teachers play a critical facilitating role, but the main task is to eventually make themselves redundant or dispensable to students' progress.

Advantages

- In this method, students do "learn to learn," so that they can meet their personal lifelong need to seek and adapt new knowledge to the challenges and problems they will encounter in the future.

They can make their present learning relevant to their future educational needs, career, and their style of learning, and they can pace their learning appropriately according to their ability to learn and understand more contemporary content. Curiosity and interest are usually the motivational factors. Because their learning is self-determined and acquired through their own "digging" or study, students become active participants and personally invest in facilitating their own learning.

- As students become active participants in the learning process, student choice facilitates motivation and interest as students create their own context and relevancy within problems and projects.

The rewards become internal and less teacher dispensed. Learning, pace, content, style, self-evaluation, and resource determination become a collaborative effort between teachers and students. Students acquire the ability to evaluate their own strengths and weaknesses, to determine their own needs, and to learn to meet those needs.

- The student and teacher share the burden to find up-to-date references or learning resources.

This includes learning how to obtain and use the resources. This model reflects the skills required for success in many career pathways and in graduate-level college or university work.

Disadvantages

- Student-centered learning creates many organizational problems.

Extensive learning resources must be available to create the least restrictive learning environment possible so that each student or student group can easily pursue his or her own educational needs.

Problems occur because of the nonlinear nature of the curriculum, which must be less structured in order to allow students to spend time using the available resources, as they feel appropriate, to be used in their own educational designs.

◪ Assessment/evaluation has to be individualized.

Each student must be evaluated within his or her own context structure. Assessment and evaluation are based around a student's own goals and mutually agreed-upon criteria. This arrangement is an advantage to the student, but may be seen as a disadvantage to the teacher. Of course, each teacher sets certain nonnegotiable goals that any school must require of its students. The student, by accepting a position in this type of learning environment, must expect that there will be a number of competencies to be mastered. Despite this, many teachers see this learning style as a messy model. Some teachers are not comfortable with the role change.

◪ This approach can create insecurity on the part of students, parents, and faculty. Changing the learner's responsibilities requires foresight and planning.

In the beginning, students worry about their ability to determine what they need to know and in what depth. Many students have learned to be passive learners and do not engage and adapt easily to this model. Many teachers cannot trust or imagine that students can learn on their own.

The student-centered approach requires maturity and discipline on the part of the students and different types of educational skills for the teachers, who must be able to facilitate, guide, and evaluate students as individual learners who are equally responsible for their own learning. These are qualities that lifelong learners must master and possess. There is no better time to develop them than when their growth can be enhanced and monitored by teachers and parents.

Any program or curriculum designers who underestimate the insecurity that students and parents feel during a change in teaching and learning strategies may be dooming their program to failure. Change and anticipating the reaction to change is a big part of implementing any new program. If students feel a threat, they are going to blame the program for their failure, and their parents will hear about it. If the parents do not call you, they will call the administration, and you will hear about it. Damage control sometimes takes care of this, but at other times, students panic and attempt to leave the new program. If the informal student leaders leave, then they will influence others.

Timing, support, and flexibility in change and innovation is important. Communication and marketing are also crucial. Parents and students fear the unknown. Be liberal in the communication of any new program, pedagogy, or curriculum, especially the expected product outcomes, assessment techniques, and expectations for students. Involving the students in creating changes makes them participants in rather than recipients of change. They do not want to feel that the program was dumped on them. Giving low grades because students do not engage in the activities or doubt your commitment to the new program can be devastating. You can plan for this. Parent and administrative support can be garnered in the same way. Involve them. Talk to colleagues. The approval of influential people bestows validity and credibility on the innovative changes. Plan for changes and do not underestimate potential problems.

Subject-Based Learning

This model is the most familiar and recognizable school organization, in which knowledge and related processes are arranged into subject areas. It is the layer-cake model of separating the sciences into isolated subjects. Because math is usually integrated into lessons and emphasized to a greater degree in chemistry and physics, these two classes are taught later in the student's high school career. Within these subject areas, learning may be organized into a hierarchy of specific, basic knowledge that builds up to more advanced or complex concepts. The goals and objectives within subject areas are for the student to gain an overall grasp of the subject, to learn the important concepts in sufficient depth, to have an understanding of the field itself, or to apply concepts from that field to future tasks. Again, this method is independent of format because subject-based learning can be individualized and self-paced. It can be teacher centered or student centered, as long as it is organized around and focused on a subject.

Advantages

- ◪ In this system, the end points or limits to student learning are defined by the subject area, as is the sequence of learning.

The extent and depth of knowledge to be acquired is more easily defined for the teacher and students. Resources for learning in one specific subject area are more easily identified and made available for student use. Each subject area has its own set of equipment and supplies located in a centralized area. Curriculum content is easily defined.

- ◪ This approach seems efficient because students apply themselves to the task of memorizing or manipulating the concepts, skills, and information within a narrow, related focus.

In teacher-centered delivery, evaluation is easily designed to sample the student's recall of specific knowledge and concepts identified through the use of convenient and well-established testing strategies.

Disadvantages

- ◪ In subject-based learning, the information acquired is not conveniently integrated with information from other disciplines or science subject areas.

There is limited context or relevancy outside of the subject area. This is especially true of mathematics and the sciences. The subject is rarely learned or used in a realistically applicable context and may limit a student's ability to organize information within his or her memory, except for the subject or course.

- ◪ Competency in connecting and integrating content in subject areas requires practice.

If cognitive connections among subjects are not actively laid down during the learning process, one cannot expect students to intuitively develop these connections when faced with problems where information from a variety of disciplines has potential applications. In a layer-cake science approach, biology, usually taught first, is often

taught without math, and teachers may gloss over biochemical principles and concepts because students have not had a chemistry class.

- ◪ Evaluation usually focuses on the subject only and the ability of
  a student to recall a narrow and limited amount of information.

Examples include a science teacher evaluating or assessing for science
content only, not for grammatical errors or mechanics, or a math
teacher presenting problems with no context to authentic applications
outside the classroom.

Problem-Based Learning

There are many philosophical models used as frameworks or structures around which to build learning experiences and perspectives.
In Elliot Eisner's (1985) book *The Educational Imagination: On Design
and Evaluation of School Programs,* five curricular orientations or philosophies are described, and the problem-based approach can combine them well.

The Development of the Cognitive Processes

This is one philosophy that stresses helping students learn to learn
by providing them with opportunities to use and strengthen the
variety of intellectual facilities or cognitive realms that they possess.
The mind is seen as a collection of independent facilities or aptitudes.
They range from the ability to infer and speculate to problem solving
and memorization—to focus on any one of them is counterproductive. Strengthening the cognitive processes allows students to cope
with today's problems as well as future problems. Acquiring information, facts, or theory, in this view, leaves the student in a poor position
to deal with future problems. What is transferred and emphasized is
not content but process. Curriculum is embedded in Bloom's taxo-
nomic model for cognitive development, which ranges from low-level
(knowledge, application) to high-level (analysis, synthesis, evaluation) cognitive function. Proficiency at each level becomes the curricular objective. Teachers build problematic situations for students
and structure them at cognitive levels that challenge students and

direct their attention to germinating cognitive processes and approaches to deal with the situation.

Personal Relevance

This model stresses the embedding of content and process with personal meaning, and it is developed site by site and classroom by classroom. In development terms, teachers design curricula in concert with students rather than from sources outside of the school. A major component supporting this orientation is that teacher-pupil planning creates personal investment. Students are viewed as individuals who require real choices or options within curricular activities to maximize the activity's potential to meet student learning needs. This model also encourages personal rapport between students and teachers. Without rapport, the teacher would not be in a position to understand the character of the student's experience.

The Technical Problem Approach

A third orientation approaches curricular planning as a technical problem, defining outcomes and setting goals, points, and styles of measurements to quantify what has to be achieved. The curriculum contains appropriate obstacles and hurdles (learning tasks) that have been formulated. This model treats learning as a product. It is an industrialized model that mandates that the curriculum be designed to stress accountability and provide evidence of educational effectiveness.

Social Adaptation and Reconstructionism

The final two orientations are social adaptation and social reconstructionism. Social adaptation can best be visualized as the school system's response to Sputnik. Schools were mobilized to produce programs that would prepare students to meet the challenges of the cold war in science and technology. Schools and programs exist to yield students who can solve societal problems and keep our society competitive globally. Social reconstructionism employs a thematic approach that uses social problems and controversial issues as curricular vehicles. The aim is not to help students adapt to society but to recognize real problems and do something about them.

Combining the Models: Problem-Based Learning

Thoughtful personal insight and careful formulation, design, and development can produce a problem-based curriculum that unifies these orientations. Cognitive process, personal relevance, technical accountability, and social orientations are all elements of the problem-based approach.

Learning from problem situations has been and continues to be a condition of human existence and survival. Clearly, problem-based learning is a basic human learning process founded in patterns of reasoning that allowed early humans to survive in their environment. Reducing this concept and approach to specific classroom practice is a natural extension of a basic human process. In this approach, the students take on problems or projects related to science subjects as a stimulus for learning in the content areas, subjects, or disciplines. In doing so, the students exercise or further develop their problem-solving skills. This approach or method of learning has two educational objectives: the acquisition of an integrated body of science knowledge, concepts, and principles related to the problem, and the development or application of problem-solving and reasoning skills.

Problem-based learning is ideally suited to student-centered and individualized learning. In a student-centered model, students may choose a problem within a larger topic. They design, develop, and modify the mode or pathway to resolution of the problem. This includes decisions on what is to be learned, which resources to seek and use, and how communication of mastery of the problem is to be presented. It also can be used in a teacher-centered approach in which the teacher can specify the problem to be addressed, the area to be studied, and the resources that are appropriate. This will develop students' problem-solving skills and involve them in active acquisition of knowledge, but they are not involved in designing or creating the problem.

Advantages

This approach provides advantages for both the acquisition of knowledge and the development of essential skills necessary in many careers. The following points summarize the positive advantages of a problem-based approach to curricular development.

- It models the way science and general learning are practiced in the real world.

Rarely does any knowledge base remain static. In most science and other professions, knowledge is very dynamic and requires contemporary understanding for optimal success in problem solving. Information, concepts, and skills learned by the student are put into memory associated with the problem. This improves recall and retention when the student faces another problem where the information is relevant.

- Problems actively integrate information into a mental framework that can be applied to new problems.

The major concepts of science can be embedded in problems in a quantitative way. Ideally, the interdisciplinary nature of science comes together with connections to other disciplines. The blending of interdisciplinary knowledge and concepts, acquired while involved in relevant problem-based curricula, creates an ideal framework for a better understanding of complex relationships in a more authentic context.

- By working with an unknown problem, the student is forced to develop problem-solving pathways and scientific reasoning skills.

Generated problematic situations at a level of analysis that students would not likely use without a teacher's guidance present open-ended challenges. These skills are universal in nature and can be applied in many other areas.

Students must define; understand and apply questioning; gather information; analyze and synthesize the data available; develop problem-solving designs; and adjust and evaluate the design in an ongoing manner. These components reflect the critical examinations that professionals face in real life.

◪ The entire range of cognitive and intellectual processes are exercised and strengthened in well-designed problems.

Success in these experiences fosters autonomy and self-reliance, which are essential qualities. Site-based, problem-based curricular development can embed a personal connection and application to the student's community or personal life. This type of facilitation connects and defines the roles of science in society. When knowledge is centered around a project or problem, the student can see the relevance of what he or she has to learn, particularly the importance of basic knowledge, which is integrated and required by the problem-solving process.

◪ An added reward to problem-based/student-centered learning is the discovery by those teachers who become comfortable with this approach that the method is enjoyable, rewarding, and a more natural way for students and teachers to collaborate.

Students and teachers become coworkers. Students, once accustomed to this learning style, become more excited and engaged, and they show a more mature behavior (they are being treated as self-determined adults). They evolve secure reasoning and learning skills and acquire a solid groundwork of basic knowledge.

Disadvantages

The following are the most commonly voiced concerns of the problem-based style of teaching and learning. They may be seen as disadvantages but can be addressed, considered, and mitigated.

◪ Creating and bringing in a new learning style may create some anxiety in learners.

Students who have been successful in other learning models will have to adjust to the new expectations that come with a change in pedagogy.

◪ The success of problem-based/student-centered learning depends on students disciplining themselves to work with unknown and possibly puzzling problems in a way that will challenge the development of their problem-solving skills and stimulate relevant self-directed learning.

The teacher must have the skills necessary to orient and guide students and design and produce or assemble problem-based learning materials.

◪ There is a perception that the problem-based learning method stresses process to the detriment of learning basic knowledge.

The understanding and retention of basic knowledge can be an outcome of this method. A balance between content and process can be built into the activity. If student expectations at the beginning of any problem-based activity include content retention as part of the assessment and evaluation process, students will need to respond to those expectations. It is up to the activity designer to build these goals and objectives into the activity and clearly communicate them to students. Clarifying goals and objectives also defines the possible areas or depth to be pursued when working with the problem. A technical understanding of a concept can be required and expected in order to fully deal with a problem. Either the complexity of the problem can drive the need to know, or assessment can be the invisible motivational force.

◪ Another concern is that teachers feel insecure with students going in so many directions; learning and teaching looks messy.

Students are not in rows, and it can be noisy. Students, if properly oriented and guided, can learn the basic curriculum in any area and to any depth or rigor. Many real-world work sites have a variety of activities going on all the time. Rarely do you find 30 people all doing the same thing. The ability to work in a loosely structured classroom can be part of the assessment process. We all need to learn to work during periods of distraction. Usually, creating short intervals between turning in work products keeps students on task. As students adjust and begin to use time well, the intervals can be increased. Younger students typically have the most problem with a less structured setting. The most important factor in the student's effective use of the problem is a clear understanding of the objectives and outcomes, both intermediate and at the end of the experience. Cooperative learning and effective use of time can become clear goals. Objectives and expectations provide both students and teachers with behavior guidelines.

◼ To some, problem-based learning seems to be an inefficient way to learn.

When confronted with a unfamiliar problem, students require considerable time to brainstorm, understand the problem, and begin to structure problem-solving pathways. There are so many important and relevant areas, side issues, and other connections that could be studied in any problem. It may seem as though an inordinate amount of time must be spent to complete the working structure of process pathways. Teachers act as tutors, guides, or coaches in this context, narrowing the focus and identifying valid pathways, objectives, and goals within the problems. In actual fact, there is little inefficiency because much of the organizational skills and knowledge become a factual groundwork for understanding other problems.

◼ Another common concern is that these methods of learning do not directly facilitate the students' ability to pass standardized tests that largely recall isolated facts and content.

In a problem-based approach, recall occurs best when students are faced with problems, not when they are faced with subject-oriented questions out of relevant context. After our formal education is completed, we rarely encounter this type of evaluation again. Problem-based evaluation requires new and different types of tools that assess and evaluate students' ability to work with problems and apply learned information to their understanding or resolution of those problems. This is the way most of us are evaluated in the world outside of the classroom, and it more clearly reflects relevant, authentic expectations for career performance.

◪ And finally, parents are not always comfortable with this style of teaching and learning.

In most cases, they did not learn this way. They need to be taught how to help their students and become comfortable with this style of teaching and learning. The effectiveness of the problem-based learning approach has been studied in various medical schools, where successful problem solving is a major goal of medical education. These studies show that it takes up to 6 months for students from more traditional programs to acclimate to this approach. This time lag can be a problem for parents, teachers, and students. Dutch medical school studies indicate comparable achievement between traditional programs and problem-based programs. Problem-based curricula appeared to provide a friendlier and more inviting educational climate. Such educational climates and environments facilitate positive attitudes, greater interest, and motivation, and they provide an effective system for the processing and retention of new information.

Summary

Teacher-based learning refers to students looking to teachers for information that may be important for their success in the class. This usually comes in a lecture format with teacher-directed activities, and generally uses recall types of evaluation. Student-based learning refers to more individualized or more self-directed study. There is no doubt that teacher-based learning is an efficient way to cover large amounts of information. Student-based learning can be facilitated best with self-study units of one type or another. It is important to see these approaches as independent of content format. The lecture can be student-based if students have input on subjects they think are important, or students may actually deliver some of the lectures. Self-study can be teacher-based if the teacher structures the units, specifies the readings and other experiences that should be undertaken, and sets the time frame and outcome objectives.

Subject-based learning can include some aspects of the problem-based approach with problems coming from a narrow content focus with little integration of information and concepts from other disciplines. Problem-based learning takes a more holistic approach.

Problem-based learning is the learning that results from the process of working toward the understanding or resolution of a problem. Curricular models designed in this way may more clearly simulate the skills required and conditions within many career paths.

Activities must guide and engage students in acquiring knowledge and concepts while also developing more universal learning protocols. Interactive project-based or problem-based learning experiences create a dynamic context for acquiring the knowledge, concepts, and processes taught in schools. Rather than learning isolated facts and procedures without direct applications, this type of learning invites and motivates students to learn to solve or create solutions to relevant problems or complete projects in a less contrived, more authentic context. Problem-based learning puts the burden of education on the student, the person most interested in his or her own educational progress. Learning activities such as these more clearly resemble career paradigms.

Knowledge derived from problem investigation creates clear relevance, importance, and significance to the understanding and management of the information and problem. In integrated interdisciplinary problems, the emphasis on each subject is related to its ability to contribute important tools and knowledge. Interdisciplinary collaboration and connections are enhanced and fostered. Students are able to evaluate resources and immediately have an opportunity to apply their knowledge.

2

The Basic Elements or Characteristics of the Problem-Based Curriculum

The structure of the problem-based curriculum is loosely summarized here (Barrows, 1985; Barrows & Tamblyn, 1980; Kaufman, 1985). These elements or characteristics define a specific curricular, problem-based learning paradigm. A well-designed problem-based curriculum contains the following basic characteristics; these should be considered guidelines rather than rules.

◪ Self-directed acquisition and the accumulation of processes, techniques, knowledge, or information bases that are

a. Organized in a more relevant and useful context. Knowledge is associated with more authentic applications and cognitive and intellectual structures, and it is acquired in such a way that it increases retention.

b. Recalled and retrieved in a real-world science context, whenever possible, as needed in discovery/investigation or to solve or explain curricular problems. Concepts and facts are mentally attached to more memorable frameworks or contexts within the selected problems.

This can be illustrated. Learning to grow and count e-coli and fecal coli bacteria from a sample of river water to assess water quality for an environmental study is much different from just going through the motions in a "canned" laboratory experience.

Another class may investigate how sea urchin food preference or a dead urchin can contaminate their seawater aquarium. Ammonia and nitrate levels go up. Students need to learn some chemistry quickly to save the tank. Simple aquarium test kits from a pet store will help them quantify the levels of the chemicals in the tank.

These two group examples illustrate how the application of textbook knowledge makes learning more relevant and interesting when it is applied to work in which the students are interested. It makes curriculum coverage harder to quantify and a little more messy, but the trade-off is worth it.

c. Easily revisited and extended through future self-directed study or applied to unfamiliar situations. Once an authentic framework for content is established, additional information can be added to the foundation. Associations to interesting problems and investigations are retained longer than a chapter test.

d. Reflective of an interdisciplinary/integrated scale of application from global to local, abstract to concrete. Carefully constructed problems connect to other subjects and disciplines, which validate and strengthen the curriculum within all areas that the problem touches.

e. Connected to a framework of personal reference, experience, or context, if possible. If the problem comes from or reflects something with which the students may have personal experience, interest goes up.

◪ A development of analytical reasoning skills (problem solving) in using the knowledge acquired

These learning experiences lead to habits of the mind needed as lifelong learners. This includes expertise in maintaining and acquiring an ongoing knowledge base as required throughout most professional careers in synthesizing solutions and products. The assumption is that competency in the problem-based approach is of future and current necessity. Saying this in a more basic way, experience in solving problems while learning is called gaining experience in the workplace. It is the difference between being "book" smart and having a more holistic, practical approach coupled with the knowledge.

- A development of self-directed learning activities and skills, including self-evaluation and self-monitoring, as well as skills in using a variety of information resources that more clearly reflect skills required outside of the classroom and school

Learning how to follow directions that are typically associated with many science laboratory experiences with linear pathways to "right" answers does not reflect the reality of many types of problems in science and other disciplines outside of school. Sometimes, there are multiple pathways within problems by which to find solutions.

- Encouragement for the development of independent and critical thinking opportunities and pathways, for groups and individual students, within problem-based learning curricula

Different components within the same problem allow specialization and avoiding head-to-head competition for the same answer or solution. Lowering the competitive level of the class may be appropriate for heterogeneous groups of students.

- Encouragement for development of opportunities for cooperative or team learning situations, which mirrors the relationship requirements in many real-world settings
- Facilitation of opportunities for the integration of past information, knowledge, and experiences within current student learning activities and the incorporation of mechanisms for acquiring, applying, updating, and adding to past knowledge. This spiraling pattern of revisiting and building on concepts is a natural way to reinforce background and knowledge base.
- Provides the student with:

    a. A motivating and exciting learning method. It allows students to produce intellectual cognitive products rather than just consume information and recall it for assessment and evaluation. Presenting the conclusion to an original open-ended study, investigation, or experiment (products) can reflect mastery of not only content but application.

b. The perception that his or her learning is relevant and is information in context with its application.

Another example begins with a student wondering what it would be like to let dry ice (CO_2) vaporize slowly around a tented plant. He knows about how plants use CO_2 and would like to know if he could enhance growth. He begins to think about what limits there might be to growth and goes back to the book. After learning the Calvin Cycle, his next question is whether plants could even handle extra CO_2. No real answers, but he is thinking and has a real need to understand the textbook.

c. A method that allows for individualized, open-ended learning opportunities. The depth and rigor of expectations for students can be made appropriate for multiple learners.

All of these characteristics are addressed simultaneously in the problem-based learning approach.

What Makes a Good, Specific Problem, Theme, or Topic?

These are the elements of any single problem, theme, or topic within the problem-based learning approach. As with problem-based learning characteristics, these should be considered guidelines, not rules.

◪ A scale of context and relevancy, from global to local, is present.

For example, global warming begins in a student's own community. Greenhouse gases are being emitted in most cities and towns. How would a rise in temperature change microclimates in the student's own community?

◪ Some aspects of the problem touch a student personally. Usually, problems within the student's own community are more interesting than general workbook-type problems.

Water quality problems turn into habitat problems that are faced by most cities and towns. Many large construction sites require environmental impact reports. Many environmental impact issues are biological in nature and can become interesting and can occur right in the student's backyard. Local fish and game people deal with many biological problems. These agencies deal with other investigations and problems outside of hunting and fishing paradigms. A local university or college may have research projects that may be adapted for the secondary classroom. With a few phone calls, you may be able to find something that could be turned into a problem-based unit.

◪ It must be a reflection of a contemporary situation or a creative and original juxtaposition of contemporary understanding and past understanding.

How do flu viruses of today behave differently from those in the past? What happens when flu hits the school's community? This is a good opportunity to teach viruses.

◪ Moral and ethical connections are built in with a balance between the emotional and more factual concrete components of the problem.

Exploitation of natural resources always contains these components. The allocation of research money toward basic research versus applied research is always interesting.

◪ Problems can be integrated and interdisciplinary, but content can emphasize a narrow range of specific subject objectives.

Math, English, and communication skills can become big parts of any biology curriculum. The world has a biological history. Human history has been shaped and molded by our biological history from containment of disease to human longevity and development of technology. Math is sometimes neglected in the layer-cake version of secondary science. Most patterns in biological research are identified, interpreted, and communicated using numbers. Math is the language of the sciences, and science can complement and validate math concepts within authentic applications in the science classroom.

◪ Connections between disciplines can fit logically together in the problem matrix or structure.

Biology is the application of chemistry and biophysics. Not much happens in biology that cannot become a chemistry lesson.

◪ Pathways are complex enough to accommodate a variety of investigative and individual learning style pathways for individuals and groups.

Open-ended problems accommodate many interest, ability, and motivational levels within the same classroom. Many examples of this notion are contained in activities within other chapters. One classic example is offering students a packet of plant seeds and asking them to construct a research project.

The complexity and sophistication of investigative questions are limited by imagination, equipment, and materials only. Testing fertilizers, soils, water, and the effects of gravity, light, temperature, and a variety of other factors can be as rigorous and complex as a student is willing to tackle. Experimental design, controls/experimental factors, and quantifying results all can be challenging.

◪ Avenues or pathways for individual interests and explorations occur within the same problem.

One water quality study involved testing water for ammonia and nitrates in local dairy farm run-off. Two students collected the water at study sites during heavy rain events. One was responsible for the laboratory tests, and one did the number crunching on a computer. In a side project, two others did bioassay testing looking at survival rates of Drosophila eggs exposed to various concentrations of these chemicals. Toxicity data supported the idea that high concentrations might cause insect larvae death in streams and creeks near the dairy. These test kits can be purchased in aquarium stores and through biological catalogs.

- Critique comes from a wider audience rather than one teacher whenever possible.

Presenting results to other classes helps put a more authentic twist on a presentation, which in turn may motivate students to set higher personal standards for performance.

- Rigor is based on the complexity of the problem and the steps necessary to respond to the problem situation, not the complexity or amount of information covered.

Sometimes, a well-planned response to a problem contains beauty in thinking and mental logic. Depth of research, overall effort, and clear presentation exhibit mastery of principles or concepts. Success is based on mastery of the problem-solving process, and basic content mastery is only part of the equation.

- Responses created and presented are based on real-world standards when possible, not just "school grade" responses and presentations.

Setting high standards for presentation and communication of results or the production of work products is important. Whatever is expected in the real world should begin to define the standard for outcomes. Real-world expectations become goals. Artists treat their work differently from secondary art students. Fingerprints and folded or wrinkled work are things you do not usually see in a professional artist's studio. Facilitating high personal expectations is an ongoing process.

The Limits of the Problem-Based
Approach in the Life Sciences

When designing, developing, or orchestrating problem-based activities in the life sciences, there are limits to what can be done due to the nature of the scientific inquiry process. Unlike other subject disciplines, problem-based, open-ended investigation and research require data collection. Collecting original data on the organismal or ecological scale requires planning, patience, and persistence, but in many cases, little equipment or complex techniques. Conversely, collecting data on the cellular or molecular level requires equipment and specialized techniques. Gathering data while working with genetic materials, proteins, DNA, and RNA is beyond the scope of many science classrooms and teachers. Authentic research on this scale is dominated by complex procedures and sometimes materials and chemicals that are not appropriate for secondary schools. There are programs that work on this scale, but learning technique dominates instructional activities. This leaves little time, for those that master the techniques necessary, to use their skills to answer original questions.

Microscopes, however, can be used to investigate certain cellular questions. Cells and their behavior, microscopic organisms and parts or portions of other living things can be observed, measured, and counted under dissection and compound microscopes. Tests and experiments can be done and data can be gathered in a problem-based context. Techniques can be learned in the context of larger, more authentic problems. Students learn the techniques out of the need to gather data or do experiments.

For science on the smaller scale, problem-based curriculum writers have to focus on other approaches. Asking students to explain, on a molecular level, how the dinosaurs in Michael Crichton's *Jurassic Park* came to life may be one way to create an authentic context and purpose for acquiring information or knowledge on the molecular level. The *Photosynthetic Human Project,* discussed later in this book, has a genetic component and may work for certain classes. Also, the *60 Minutes Project* touches on the use of molecular and biochemical techniques in a larger context and relevancy.

Writing on ethical questions involving genetic engineering requires the writer to understand the workings of science on this level. When debating a particular standpoint, engaging in an argument with factual background and understanding always is more convincing than arguing from emotion with little background or understanding.

Scientific knowledge, process, and technique can be acquired in context in other applications and context paradigms. Lawyers arguing DNA evidence may never have put on a lab coat, but they have to know their science to successfully negotiate the courtroom. Reporters and writers must learn the latest scientific techniques on the smallest scale to accurately complete a story or a science fiction book. The scope and scale of some science topics can be experienced in a context other than a true investigation.

For those with the interest and motivation, many repositories of genetic data, such as GenBank, offer sources of free data. Nucleotide sequences are available that contain complex patterns of information. GenBank contains data on DNA, RNA, and proteins that are complex but can be reduced in complexity while retaining the same patterns. Most scientific journal articles and research investigations on the molecular level donate their genetic data to these repositories. This allows others to look at the exact data that the researchers used in their study. In other words, you do not have to generate it yourself. Data sets at this scale of science research are indeed large, and in many cases, sophisticated computer programs must perform the analysis. Again, with patience, a teacher could use these studies to create smaller, less complex data sets for students to analyze for information and patterns.

3

Science Without the Mess
Where to Find Investigations and Sources of Data

The following are potential sources of problem-based activities for individual students and student groups. All general areas contain the characteristics and elements of problem-based curricula. Problems and projects can be designed and developed to include the basic elements of a problem described in the previous chapter.

Outcomes, assessment and evaluation criteria, and both intermediate and completion goals and objectives can be created and developed collaboratively. For example, a recurring question that comes up with students is the nutritional quality of cafeteria or snack bar food. Nutritional chemistry is a theme rich in content, and an investigation and analysis of snack and lunch food is a great vehicle for the content in a few biology book chapters. It could include a class or only a few students. The experimental design would be collaborative because the class would be expected to brainstorm and create the methods for gathering data.

If heavy content knowledge within the problem is your goal, build in the expectations from the beginning. Nutrition has its roots in biochemistry. The topic can serve as a rigorous venture into the complex chemistry of foods and digestion, or it can become a reason to learn to be a better consumer and assume basic control over one's personal diet. In the cafeteria example, students can generate a tremendous amount of data quickly. Keeping track of what the "average" student eats at break and analyzing it for nutritional content is loaded with quantitative opportunities. Methods, standards of mastery, and demonstration of mastery also can be formulated with students.

How much information is needed to draw accurate conclusions? You could break the study into categories, such as girls, boys, and a variety of ages. Many options for depth and breadth of investigation and self-discovery exist. They range from knowing specific facts and processes supporting larger general science concepts to a more holistic general understanding. Expectations for depth of information source or reference research can be established. Final expectations for the delivery and communication of mastery can range from a formal expository scientific paper, oral presentations, debate, or portfolio to a combination of one or more. Data and results can be quantified and manipulated with a variety of computer programs or mathematical processes. Students will have to experience the process of doing science.

Nutritional Facts

A 10th-grade football player wants to develop a pregame diet to optimize athletic performance. What would he need to know and learn to produce a working diet? A student group of vegetarians is convinced that food from the snack bar is junk food. They want to write a letter to the principal. What would they have to know to create a relevant, informed argument? A concerned animal lover wants to know how dogs and cats can live off meat products alone and wonders if there is a better diet for them. The nutritionally related questions are endless. They relate personally to students and are loaded in scientific content, principles, and concepts that can be acquired through investigation. The questions are interdisciplinary in nature and integrate the sciences. The best part is that students can answer these original questions with little or no equipment or materials.

Data, in the form of nutritional values and ingredients, are present on all food products. This is a source of information that is presented in a quantified manner using biochemical vocabulary. Serving sizes and caloric values are included within a metric measurement format. The testing, production, and generation of this information is well beyond the capacity of high school laboratories, yet it is free and available to you and your students. The problem-based pathways are endless. These data can be used to work through a limitless number of questions and problems. Students can produce new product outcomes and generate knowledge in the process. Creating problems and projects using these data connect to many science concepts from basic biochemistry—usually an early chapter in most biology books—to human ecology and environmental land use.

There are many inexpensive diet and calorie-counting books on the market. If you have a computer, there are diet programs available that do a quantitative analysis of any meal, diet, or menu you want. Results are instant. Some of the better programs have a database of information on name-brand fast food, most have printing capacity. With a spreadsheet and graphing program, more analysis can be done.

Sports and Athletics:
Performance Data

School sports teams offer a source of information and data on fitness and training techniques, injury/recovery data, and athletic performance. Topics include exercise physiology, anatomy, and health. For certain groups of students, the scientific basis of exercise physiology and athletic performance can become a perfect vehicle to learn the science behind these topics in a relevant and interesting way. In reviewing two college textbooks, *Exercise Physiology—Human Bioenergetics and Its Application* and *Fundamentals of Human Performance*, both by George A. Brooks and Thomas D. Fahey, it became clear that you could create very complex problem-based learning modules within the sports and athletic theme. These modules not only covered the subject conceptually but in as much detail as any college biology book. You could create many problems and teach a great deal of content within this theme. The sports teams become the "lab rats," so to speak.

The following study was actually created by students. As a small group project, three students wanted to find out who were the fittest athletes in the school. Were the most successful athletes also the fittest? They had to do a good bit of work defining what "fitness" really is and how it is achieved, and then they had to design ways to measure it. They measured pulse rates and lung capacities under stress. They measured endurance on stair machines and exercise cycles. They found safe ways to measure strength and flexibility. They got almost too much data. As varsity athletes themselves, metabolism, ventilation, cardiovascular dynamics, and conditioning became personally relevant.

A second group worked on injury statistics and recovery times. They visited doctors and rehabilitation facilities. One actually witnessed a knee surgery. Again, the problem started with a question on which sports were the safest and which contributed to injuries. From that starting point, they moved into a variety of other topics that offered reasons for learning content and good research techniques.

For references, look to local college bookstores offering sports medicine, athletic training, or performance nutrition textbooks. Many biology supply catalogs offer inexpensive devices for gathering data. I gave out stopwatches, blood pressure devices, pulse rate monitors, and other tools that are safe and easy to use with little training. The mathematical potential here is unlimited for statistical analysis and validity testing.

In all cases, the students are engaged in doing science, creating new information and acquiring knowledge and curricular content.

Scientific Surveys

A possible explanation for the origin of life is discussed in class, which leads to a discussion involving when life begins. Some argue at the moment of birth and others at the moment of conception. A group wants to survey fellow students' opinions. One student offers to create a number of questions to ask 20 other students to identify attitudes regarding this issue. She wants help in defining possible answer options. How would she create a valid scientific survey with a wide range of student backgrounds and opinions? Should they ask just females or males, seniors or juniors?

Do boys or girls, seniors or freshmen, grow the most over a school year? Both heights and weights can be surveyed, measured, and remeasured to look at patterns of growth in student populations. Certain genetic traits can be researched in student populations. There are many questions and learning opportunities that can be researched by gathering data within student populations.

Creating scientific surveys can intertwine and wrap around many scientific concepts, themes, and topics. Designing a survey gives students the opportunity to build and investigate research design, gather data, analyze results, and draw conclusions. Once completed, bias, reliability, and validity can be examined and explored.

4

Big Nonlaboratory Problems (But Could Include Laboratory Experiences!)

◤ These problems are appropriate for a whole class and may be expanded to larger groups. Laboratory activities can be included in some of these projects; however, they would be included by students as part of the working of the problem. They naturally integrate many science disciplines but focus on biological concepts and principles. Some have a historical component and create a relevant context for exploring the progress of scientific thinking and application. Process and organizational activities become exercises in planning and self-direction. The teacher's role and the complexity of the work itself may vary as necessary for the individual nature of the class. The overall flow is purposely flexible, with many possible pathways to allow for individual students' interests.

Logistically, the rigor, depth, scope, and timetable can be agreed upon and reinforced during the project. They can be fit into evaluation periods or within other constraints. They can become the focus of the class or a sideline. Each project can be tailored to the needs of the teacher in curriculum design and coverage requirements. Built-in, complementary parallels to textbook content enhance the validity of both the project and the book information. Each has the potential to provide an applied math component. These problems are presented in a general format and are intended to be modified as needed. Each sample project contains a partial breakdown of the project specialty areas, tasks, and expectations. These are not meant to be rigid, however. Selected subtopic group areas within sample projects contain hypothetical examples of how student organization, management, and the overall process might progress.

Finding the Resources

During the time that I have been presenting these projects, I have collected a number of references, books, Internet sites, CD-ROM programs, and other support material for each problem. I save and bind copies of previous student responses to the assignments. I also find locations or people that are user friendly for students. These may be contacts in the local Department of Health, a dive shop, and so on. Finding resources will be the biggest challenge and problem you have when implementing some of these projects. I find that I would rather spend money on references now than on new textbooks, and I pay more attention to the school library and make purchasing suggestions that support these activities. You may want to send away for materials, spend time on the Internet, or make contacts well ahead of time. If students are able to see previous work, they usually use the better work as a gauge or standard for their own performance. This may work for you. You will have to decide what you want them to see as examples.

We want students to produce original responses to the challenges of the problems. It may be relevant to talk about ethics in writing original work before starting. You don't want to read and assess rewrites of the references.

Celebrate Darwin

To the Teacher

This project or problem naturally integrates earth, biological, and historical sciences. It introduces students to the principles and concepts of Darwinian evolution and natural selection. It also is a good way to examine how science was done at that time and how it has changed. This can set the stage for further, more current studies in many biological areas. The project is easily modified to emphasize individual teachers' needs, and it makes an excellent integrated science problem. A math, history, and geography component could be added, or a class from another discipline could be incorporated. Later activities could include a look at how today's understanding of genetics and molecular biology has added to our understanding of the principles and concepts that Darwin proposed.

What?

An anniversary trip is planned by England to celebrate and commemorate Charles Darwin's contribution to science. Organizers plan to retrace Darwin's trip that inspired his thinking on a variety of subjects. Organizers have asked that U.S. students participate in the trip. A replica of the original HMS *Beagle* will be used.

Why?

This school's students are highly regarded for their experience in evaluating, investigating, and solving real-world problems, and they have been chosen and invited to participate in all aspects of the trip. Although many students will contribute to the planning, only 10 students from this school will make the actual trip. Students are being asked to research the route, sailing requirements, and logistics for a ship of this type with a crew of 50 student sailors and biologists. Student biologists will inventory what Darwin saw and documented at the various locations that the original HMS *Beagle* visited and compare this information with what biologists find today.

Our 10 representatives will need the background necessary to assume active leadership roles within the crew. Student teams will complete the required preliminary research to support the traveling team.

When?

The organizers are interested in turning over the entire trip to students. Can students solve the technical, logistical, and social problems of such a trip? Time is of the essence. The window of opportunity for a trip such as this is limited by a number of factors. Students will need to minimize risk and limit expense. Sailing the world's oceans is very different from motorized ship travel. Knowledge of weather, climate, navigation, and ocean currents is needed to optimize trip efficiency. Departure and arrival times from all locations need to be determined. Students are asked to develop a timetable for travel that minimizes risk and expense over the same route Darwin sailed.

How?

Students will be placed randomly on planning research teams to address major research categories. Attention will be given to balancing the teams in terms of gender. Each team will be given an opportunity to bid for a specialty area. Once an area is identified, students within each team will select a first and second choice regarding areas of concern within the specialty area. Specialty areas are as follows.

A. Navigation, Weather, and Ocean Sailing Logistics and Timetable

Outcomes must include an authentic plan for implementation based on the real-world conditions for a trip such as this. Only a few ships in the history of science are as famous as the HMS *Beagle*. This is a case study in the actual process of exploration, with all the dangers of navigation, careful surveys, difficulties of finding supplies, shipboard tensions, accidents, the routine plodding from port to port, and seagoing skills involved.

The following is a typical group and subtopic Category A example of project management. All of this may vary within the nature of an individual class, and the complexity and rigor may vary with the resources available. The following is a hypothetical model of how a piece of a project like this could go.

A group containing eight students for Category A came together to divide the duties within their specialty area. None had the background necessary to define specific work or outcomes that the project required, so it was agreed that they would work in pairs to gather resources to begin to acquire a background. Each pair would target a specific location such as the public and school libraries, the Internet, the local community college library, and a natural history museum. The group picked a student coordinator to ask the teacher for more time and suggestions. It was agreed that the resources should be brought to class the following problem work day. Each student would summarize his or her resources to share with the group. If possible, each would also copy useful information for the others.

The group members came back together and found, after brainstorming, that the larger tasks involved the earth science topics of weather; climate patterns and ocean currents for navigation; crew outfitting for clothes and food; and identification of resources, both natural and manmade, or services that were available at landfalls along the trip's route. This includes an assessment of how user friendly the country might be. Political climate is another consideration. The Internet has a CIA gopher and web site with "country profiles" available. These profiles give you the most current information on a large number of world countries. Some are very extensive. Also, the U.S. State Department has information in the form of travel advisories. Many people send for these, but they are instantly available on the Internet. The advisories cover many risks and concerns that travelers face when visiting various foreign countries. Students found these useful because they did not want to put the crew in the middle of a civil war.

A smaller group used the log or a detailed book account of the trip of the *Beagle* for a timetable and compared the timetable with the seasons of the year in those visited areas. Unlike the *Beagle*, their ship would be equipped with a diesel engine. This would give them a little more flexibility in travel times, but the main mode of power was to be the wind. The team would try to come up with optimum sailing conditions for the areas and seasons of travel.

A second, smaller group would look at outfitting the participants with proper clothing, food, and supplies for completing the types of research in which the participants would be involved. They would need to work closely with others to learn about weather expectations (they may need to research this themselves) and begin to decide what kinds of clothing would be necessary and to estimate the food and medical needs of the trip. This work also includes the necessary scientific supplies and equipment as required by the researchers based on what researchers on the *Beagle* did. A supply manifesto and inventory in addition to a general explanation, validation, and justification for the equipment and supplies would be required. Space is very limited, so mistakes could threaten the trip.

The final pair of students would decide what type of navigation strategy to employ. Celestial navigation is what the *Beagle* used. However, for safety reasons, it was agreed to use more modern electronic navigation. This group's task was to outfit the ship with appropriate electronics and to learn how to use the equipment in a hypothetical way. It was also agreed that they would review celestial navigation techniques to share with other team members. Resource acquisition would begin with one student contacting some relatives who owned a sailboat and had sailing experience. The other would call a local Coast Guard station to ask for advice. One student with Internet access would gather information on any availible weather and sea data on the trip route and look for satellite information from the National Aeronautical and Space Administration and the National Oceanic and Atmospheric Administration. This group would need to work closely with the earth science group to anticipate the route and sea conditions.

Once the general goals for each group were set, the teachers and the students within each research category collaborated on expected outcomes and presentations. It was agreed that each member of a team would keep an individual portfolio with notes, references, and other support material as part of a packet of materials to be assessed at the end of the project. In addition, each group would write a mission statement describing what they were responsible for researching and producing. This would document individual effort and contribution in addition to the team and groups. It was also agreed that each small specialty group would share a preliminary report 2 weeks into the project. This would serve to clear up last-minute issues and keep all group members on track for the final outcomes or reports.

The final outcomes will be written reports compiled by team leaders and put into a book or manual format. The team leaders will decide the page and text formats for all. It was also noted that authenticity was the key criterion. The more authentic the plan and the more current the information and other data, the more valid the final report. A rubric was developed with the collaboration of the teacher and group leaders to define individual evaluation of the individual portfolio. Each student would be assessed on his or her individual contribution in addition to the evaluation of group products (see assessment and evaluation at the end of the project).

B. Darwin's Activities and Background

This is almost a role-playing activity. A small group of students would assume Darwin's role on the trip. They would become the historians and Darwinian experts. Students would need to acquire a detailed understanding and interpretation of the work that Darwin did on the trip, which would allow them to communicate in a way to "teach" and inform the other trip participants. This information is necessary to provide concrete points for biologists to compare to the current survey. Many of Darwin's activities on the trip contributed to certain biological concepts that he wrote about after the trip. This also needs to become part of the information for communication. To summarize their duties:

- They become experts on every aspect of Darwin's research.
- They work with biologists to compare and contrast Darwin's description and interpretation of the biology of the sites visited on the trip and the biology of those sites today.

C. The Way It Is Today

This specialty area requires students to do some preliminary research into how student biologists would attempt to duplicate the activities in which Darwin participated. Unlike Darwin's time, there is information available on the various activities in which Darwin participated and on the locations Darwin visited. Site reports are needed to aid and guide student researchers. Each location requires as much current information as possible. What would biologists need to do at each location, and what do we already know? To summarize the task, this group is responsible for

- Assessing the biological conditions of Darwin's study sites today
- Comparing and contrasting the changes to those sites
- Creating a detailed summary of the changes for future groups to use as baseline data

D. Organization, Leadership, and Production

A small number of students should be responsible for project continuity, communication, guidance, assembly, and production of the final outcome or product. These students set the tone for defining the professional presentation requirements of the project. They will act as

◪ *Guidance counselors.* As leaders, this group will work with others groups to plan, coordinate, collaborate, and facilitate communication between groups. They will work to manage the project process.

◪ *Editors.* As editors, they will be responsible for reviewing the science.

◪ *Content, technical material, and information generalists.* The information that groups produce should make sense and reflect a quality and validity of authentic effort. Ideally, each member of this group would become an expert in one of the specialty groups.

◪ *Production experts.* Standardization of formats and the formation and communication of timelines will be part of their duties and help the project stay on track.

The leadership component within any class can be fragile. Bossy, aggressive, or pushy students can be a problem when working with other students. Each class has its own chemistry. In certain classes, these project leaders can be appointed, and in others, they can be selected by the other students. Their leadership roles become balanced between the teacher-centered curriculum and student-centered, more self-directed learning. It really depends on the specific class and maturity of the learners.

Available resources need to be considered when forming groups. Resource acquisition is based mainly on the motivation of the students. Some will really dig and others will want the resources handed to them. There are many television programs on many aspects of Darwin's life. A good tape library supplements books and other more traditional resources. Living near a college or university library offers students the opportunity to expand search strategies. It is important to have a variety of avenues available for students who are less than self-directed. Frustrations associated with lack of resources can sink the most valuable curriculum. For this reason, mixing motivated and less motivated students within groups may solve some resource problems before they start, or a group of less motiviated students could give you a chance to concentrate your teaching.

Tasks

The beginning of the project may look something like the following scenario for this group. The first few days will be spent organizing group and individual goals and gathering information. Students will need to break down tasks into identifiable categories or subgroups. Once these categories are identified, students are free to pick specific tasks for research. Some tasks may require more than one student. The first tasks for the subgroups will be to brainstorm, design, and develop strategies, goals and outcomes for their areas and some justification for their outcome's contribution to the project. An outline of these tasks will be used to construct a timetable for completion of group and individual work. Specialty areas will have to maintain team connections to support the total group effort. A leadership structure will be necessary to aid organization and communication.

Research Strategies

Team members must cooperate to identify and obtain resources in a timely manner. Documentation of sources is essential. A timetable may be necessary to keep the work on schedule. Intermediate outcomes are identified and required by instructors in addition to the final product.

The three large groups and the smaller leadership group will submit a final product. Each group's product, with each contributor's specialty outcomes, information, or documentation, will be placed in a manual and become part of a package to accompany the 10 representative students.

Individual Student Outcomes

Portfolios

1. Each student will maintain an individual portfolio documenting all work done for the journey. This applies to smaller subgroups as well. We want to see your individual contribution.

2. Each individual or subgroup will complete a report of its findings, and the information will be organized and placed in the specialty team document.

3. Each team must agree on a common style guide so that fonts, type size, format, and organization of the final document are consistent.

4. Individuals will provide evidence of mastery of content and process, overall understanding, contribution, and effort with their completed portfolios. This, in addition to intermediate outcomes, will contribute to individual assessment.

5. Individual portfolios must contain:

   a. A cover letter introducing yourself and your role and documenting your personal experiences and contributions

   b. A preliminary report, mission statement, and bibliography (standard format)

   c. Notes

   d. Copies of research documents

   e. Collected material such as newspaper articles, travel brochures, and so on

   f. Specific individual contributions to the team rally manual

   g. Copies of intermediate outcomes

   h. A self-evaluation with grade expectation

Evaluation

Individual portfolios will be worth 80% of the total grade. The team's manual equals 20% of the grade. The individual portfolio grade will be based on the following:

70% portfolio: organization, completeness, depth of research, and presentation quality

20% teamwork: collegiality and team support

10% involvement: rigor, depth, and degree or level of effort.

Challenger Deep—
To the Bottom of the Mariana's Trench

To the Teacher

This problem offers opportunities to integrate other sciences into the investigation. The main biological focus is on marine sciences. The overall goal is to investigate, interpret, survey, and inventory life at the site and identify sources of potential damage to the environment. Planning for the diving and submersible requirements of the project offer unique opportunities to integrate physics, chemistry, engineering, and human biology. Gas laws, pressure, and many other physical science concepts and how they affect human performance while diving underwater create many opportunities to explore science concepts in context with real applications. Investigations could include laboratory demonstrations of the applied principles and concepts required in underwater activities. Also, marine geology and plate tectonics could become investigative pathways. Similar logistical problems of travel exist here as within the previous problem. Mini-problems include traveling to the site and locating the actual research location. The local dive shop or an oceanographic research facility may provide important resources.

What?

The Challenger Deep is the bottom of an ocean trench system described as the deepest point in all the oceans. It lies within the Mariana's Trench system. Its isolation makes it an interesting control site for baseline studies in ocean environmental conditions. These studies will serve as points of comparison for later studies. Patterns of changes would be easier to identify with ongoing periodic collections of oceanic and marine biological data. It could be considered the most isolated water environment. This includes establishing a permanent underwater research site or colony. An international group of researchers has asked that a group of students, representing various countries, be involved in all aspects of planning, implementing, and participating in an ocean investigation and survey for environmental health and conditions within the targeted area.

When?

The timing and length of the trip is to be determined by the students. Overall goals need to include minimizing trip risks and expenses. Weather and sea conditions need to be considered.

Why?

This school's students are highly regarded for their experience in evaluating, investigating, and solving real-world problems, and they have been chosen and invited to participate in all aspects of the trip. Although many students will contribute to the planning, only 10 students from this school will make the actual trip. An overall inclusive plan for the entire trip will be produced by students for consideration and evaluation by expedition organizers. We have a limited amount of time to plan. Many technical and logistical problems need to be identified and solved. Researchers are interested in establishing a permanent underwater research facility for biologists. Could this be done?

How?

Students will be placed randomly on planning research teams to address major research categories. Attention will be given to balancing the teams in terms of gender. Each team will be given an opportunity to bid for a specialty area. Once an area is identified, students within each team will select a first and second choice regarding areas of concern within the specialty area. Specialty areas are as follows.

A. Trip Logistics and Habitat Design

How do we get there? A team of 150 individuals, researchers, technical people, and crew will be needed at the site. When do we go and how long can we stay? What do we need to have or take to complete the overall goals? Tasks for which this team will assume responsibility include the following:

- ◪ Choosing the ship and specialized equipment to outfit the ship and meet the needs of researchers and divers
- ◪ Traveling and navigating to and at the site

- Transporting the full team to the site and maintaining a position above the site
- Monitoring oceanic and atmospheric conditions during the trip. Weather, climate, and currents at the site need to be understood and planned for.
- Developing a plan for the location of the habitat and considering its potential location and its value as a research site, because not all sites would maximize research potential. What is the maximum deep that could be maintained?
- Creating plans for submersibles and the creation of the habitat. Again, consider the deep, construction of the habitat, and accessibility.
- Understanding the geology and geography of the sides and bottom of the trench itself

This section can be handled much like the Darwin Team "A" example.

B. Biological Research Team

The overall goal of the trip is to establish baseline standardized research methods in order to gather biome and environmental data. The purpose is to compare the data to information gained from future surveys to assess changes to the biological communities at the site. How will we complete a baseline survey considering the conditions at the site? What is possible, and what are the limits? Students on this team will be responsible for all aspects of investigation planning, experiments, sampling, trapping, data collection, and interpretation. Researchers will be required to work closely with dive teams to understand what is possible and to understand the limits of underwater work. Environmental water chemistry may be included as a research goal within this investigation.

The following is an example of how the team or group and topic "B" might be managed.

This group is composed of student scientists and researchers with goals much like anyone doing field research, but they will need to learn marine biology and physical oceanography to do their investigations. This is really an applied environmental science problem. The biological habitats are the various water layers, the sea bottom, the sea surface, and the atmosphere above the trench. Many field studies are done in two-dimensional grids. This is a three-dimensional study because the depth of the ocean drops off many hundreds of feet, creating habitat strata, or layers of life and conditions that may be very different within a few feet of each other. The same species or conditions could be isolated in a narrow area of the ocean or trench. They will also have to work closely with divers to begin an authentic plan for data collection teams. The larger group felt it could logically break into two groups: physical oceanography and marine biology. Each smaller group would be broken down further into the following smaller working teams. These categories are arbitrary in that each case is individual and may be brainstormed differently, and each group may propose a different model.

Group and Topic Example: Physical Oceanography Researcher Topics

Each category in the physical science area would need to be researched for existing data and information that could possibly be available for the area, the equipment methods, and the experimental designs that would be needed to collect additional data at various depths and areas of the study. This information would be necessary in the creation of sampling techniques appropriate for the target organisms or physical data. The effort should be coordinated with the dive team to create authentic methods for collection of samples and data. Some information could be collected by divers, but because of the depth, other information may have to be collected with sophisticated equipment.

Physical Science Topics

- *Water:* temperature, salinity levels, oxygen and other gas levels, pH, available light
- *Benthic or bottom sediments:* Collection strategies, soil and sediment chemistry; this includes both the bottom and sides of the trenches
- *Atmosphere:* General weather and climate information
- *Manmade chemicals and other materials:* Anything that can be attributed to man and considered pollution

Marine Biology Research Topics

Sampling techniques, in addition to experimental design and collection methods, would be part of this group's responsibility. All specialty areas would be required to develop the basic knowledge necessary to design the required studies and sample techniques for the organisms typically found in the area. Most of their work would be collecting samples and creating a quantitative log of the numbers and types of organisms in the study areas. How big would the sample area need to be? What would the data sheets look like and how would data be organized? The biological group would be broken into these three areas:

- *Benthic, bottom, or sedimentary organisms:* Living things on and in the bottom sediments, and collection and preservation methods
- *Planktonic organisms:* Weak or nonswimming organisms and sampling methods
- *Nonplanktonic organisms:* Swimming organisms and collection methods; in addition, a small subgroup would be established to survey the ocean surface biology

Assessment and Evaluation

Assessment and evaluation management, as with the other projects, is a collaborative effort. Both the teacher and the students work to define outcomes. These would vary according to the available resources and the character of the individual class. However, it could look much like the techniques used in the *Celebrate Darwin* problem.

C. *Dive Teams*

The dive team will be considered specialists and research technicians. They will aid biologists in all aspects of research and work closely with biologists in experimental design. There are many limits to research in this setting, and the dive team will need to become experts in identifying the limits of this type of work. All diving equipment, including submersibles and other underwater logistical problems, will be the responsibility of the dive team. An authentic dive plan for implementing an investigation such as this will be the overall outcome. This includes identifying and budgeting for the equipment requirements. Dive biology topics that will need to be considered include

- A plan for the physiological, biological, and medical aspects of diving safety. Students must understand diving at these depths and create a dive plan to minimize risk. They must know what happens to divers at these depths.
- Gas mixtures and dive equipment
- A plan for dive times, and minimun and maximum dive times at depth
- Backup medical plan for accidents and injuries. This would include nitrogen narcosis.

D. Organization, Leadership, and Production

A small number of students will be responsible for project continuity, communication, guidance, assembly, and production of the final outcome or product. These students set the tone for defining the professional presentation requirements of the project. They will act as

- *Guidance counselors.* As leaders, this group will work with other groups to plan, coordinate, collaborate, and facilitate communication between groups. They will work to manage the project process.
- *Editors.* As editors, they will be responsible for reviewing the science.
- *Content, technical material, and information generalists.* The information that groups produce should make sense and reflect a quality and validity of authentic effort. Ideally, each member of this group would become an expert in one of the specialty groups.
- *Production experts.* Standardization of formats and the formation and communication of timelines will be part of their duties and help the project stay on track.

The beginning of the project may look something like the following scenario for this group. The first few days will be spent organizing group and individual goals and gathering information. Students will need to break down tasks into identifiable categories or subgroups. Once these categories are identified, students will be free to pick specific tasks for research. Some tasks may require more than one student. The first tasks for subgroups will be to brainstorm, design, and develop strategies, goals, and outcomes for their areas and some justification for their outcome's contribution to the project. An outline of these tasks will be used to construct a timetable for completion of group and individual work. Specialty areas will have to maintain team connections to support the total group effort. A leadership structure will be necessary to aid organization and communication.

Research Strategies

Team members must cooperate to identify and obtain resources in a timely manner. Documentation of sources is essential. A timetable may be necessary to keep the work on schedule. Intermediate outcomes are identified and required by instructors in addition to the final product.

The four large groups will submit a final product. Each group's product, with each contributor's specialty outcomes, information, or documentation, will be placed in a manual form and become part of a package to accompany the 10 representative students.

Individual Student Outcomes

Portfolios

1. Each student must maintain an individual portfolio documenting all work done for the journey. This applies to smaller subgroups as well. We want to see your individual contribution.
2. Each individual or subgroup must complete a report of its findings, and the information will be organized and placed in the specialty team document.
3. Each team must agree on a common style guide so that fonts, type size, format, and organization of the final document are consistent.
4. Individuals must provide evidence of mastery of content and process, overall understanding, contribution, and effort with

their completed portfolios. This, in addition to intermediate outcomes, will contribute to individual assessment.

5. Individual portfolios must contain:

   a. A cover letter introducing yourself and your role and documenting your personal experiences and contributions
   b. A bibliography (standard format)
   c. Notes
   d. Copies of research documents
   e. Collected material such as newspaper articles, travel brochures, and so on
   f. Specific individual contributions to the team rally manual
   g. Copies of intermediate outcomes
   h. A self-evaluation with grade expectation

Evaluation

Individual portfolios will be worth 80% of the total grade. The team's manual equals 20% of the grade. The individual portfolio grade will be based on the following:

70% portfolio: organization, completeness, depth of research, and presentation quality
20% teamwork: collegiality and team support
10% involvement: rigor, depth, and degree or level of effort

Trip to Mount Everest

To the Teacher

As with most of these problems, this one offers opportunities to integrate other sciences into this investigation. The main biological focus is on the medical sciences: mountain medicine, human performance, and nutrition and diet. This project could be integrated with other disciplines because the region has some unique history and politics. The overall goal is to investigate, interpret, survey, and inventory life at the site and identify sources of potential damage to the environment. However, this does not have to be the focus. Planning for an expedition such as this offers unique opportunities to integrate the earth sciences (meteorology, geology) with human biology, ecology, and environmental sciences. Science concepts and how they affect human performance during the trip create many opportunities to explore science concepts in context with real applications. Planning a survey of mountain biology can be a side issue, or it can play a more prominent role. Investigations could include laboratory demonstrations of the applied principles and concepts required in high altitude activities. Also, geology and plate tectonics could become investigative pathways. Similar logistical problems of travel exist here as within the previous problem. Mini-problems include traveling to the site and locating the actual research location.

The *Celebrate Darwin* and *Challenger Deep* projects are similiar to this one. All the student categories can be modified to best use local resources and the teacher's expertise. The individual and group task details as well as assessment and evaluation techniques can be customized for the classroom application and teaching timetable. Each major problem can be embedded with and complement textbook content if the teacher desires.

What?

Mount Everest is the highest mountain in the world and is located in the Himalaya mountain range. Access is through a number of different countries, and the politics of getting a climbing permit and mounting an expedition of this type present interesting problems. It could be considered the most isolated terrestrial environment. What lives there and how humans have affected this environment are the reasons for the trip. Its isolation makes it a potential control site for baseline studies in terrestrial high altitude environmental conditions. The goal is to establish methods for surveying life forms, gathering data, and completing a study. This information will serve as a point of comparison for future research. This project includes establishing a permanent mountain research site or colony. An international group of researchers has asked that a group of students, representing various countries, be involved in all aspects of planning, implementing, and participating in this investigation and survey for environmental health and conditions within the targeted area.

When?

The timing and length of the trip are to be determined by the students. Overall goals need to include minimizing trip risks and expenses. Weather and mountain conditions need to be considered.

Why?

This school's students are highly regarded for their experience in evaluating, investigating, and solving real-world problems, and they have been chosen and invited to participate in all aspects of the trip. Although many students will contribute to the planning, only 10 students from this school will make the actual trip. An overall inclusive plan for the entire trip will be produced by students for consideration and evaluation by expedition organizers. We have a limited amount of time to plan, and many technical and logistical problems need to be identified and solved. Researchers are interested in establishing a permanent high altitude research facility for biologists. Could this be done?

How?

Students will be placed randomly on planning research teams to address the major categories. Attention will be given to balancing the teams in terms of gender. Each team will be given an opportunity to bid for a specialty area. Once an area is identified, students within each team will select a first and second choice regarding areas of concern within the specialty area. Specialty areas are as follows.

A. Trip Logistics

What does it take to get a team on the mountain? How much money will be needed? What travel and bureaucratic arrangements need to be made? This team will have to work closely with the medical and diet groups because they create the nutritional requirements.

B. Medical Team

Mountain travel at high altitude requires careful planning for medical emergencies unique to this type of travel. What happens to the body at high altitude? This team will establish pre-expedition training and create physical standards for those considering making the trip. This group will be responsible for all aspects of the health, medical needs, and fitness of the expedition team. They will also establish medical protocols for any potential problems, and they will identify and collect necessary medical supplies. A medical manual for each expedition member needs to be created to instruct members about the health and dietary issues that may come up on the journey. These should include a list of symptoms—both danger signs and normal changes associated with mountain travel—that individuals might encounter.

C. Nutrition

A trip such as this is a long-term athletic event. The dietary requirements for the conditions anticipated on this trip are unique. Food also means weight. All of it will have to be carried. Clean water is another problem that will need to be considered and addressed by this group. This group should develop an all-inclusive plan to meet the dietary and nutritional requirements of the entire team during the trip.

D. Project Production

A few individuals will be responsible for coordination of the production of the final product of this project. They are free to define the mechanical and stylistic standards for the project and set production deadlines. The documents are expected to be as professional as possible. Pagemaker experience helps.

Timetable for Production of Materials

As with the past project, all final assessment and evaluation will be based on the "working" portfolio. Keep copies of all work for the final portfolio. Remember, in this simulation, authentic means "like it is really going to happen."

1. End of the Period

    a. A list of team members and individual job descriptions within the major categories is required. What are you going to do and contribute to the larger group? Be specific and detailed. Who is going to coordinate and compile individual contributions? Keep one copy for yourselves.

b. The production team will begin to set writing guidelines and share them with the group.

2. *End of the Week*

a. Turn in a copy of your bibliography and working notes. Complete your "Learning Log" for the week describing what you have done, what you need to do, and general reflections, problems, and so on.

b. All groups and individuals should be prepared for a class progress discussion.

3. *End of Second Week*

a. Groups and individuals should turn in an update to your bibliography, "Learning Log," and outline of your part of the contribution.

4. *End of the Third Week*

a. A group rough draft is due. Individuals with each group will present a "work in progress" oral report. Be prepared to outline your group's work, problems, and so on.

b. Individuals turn in their rough drafts, additional notes, "Learning Logs," and notes.

5. *End of Fourth Week*

a. Final draft is due for editing and revision by production team. There may be adjustments and changes required. A Monday meeting will discuss expectations for the final outcomes.

6. *End of Fifth Week*

a. Final products are due and the trip planning will be shared with another class.

Evaluation

Individual portfolios will be worth 80% of the total grade. The team's final contribution equals 20%. The individual portfolio grade will be based on the following:

70% portfolio: organization, completeness, depth of research, and presentation quality
20% teamwork: collegiality and team support
10% involvement: rigor, depth, and degree or level of effort

The *60 Minutes* Project

To the Teacher

The problem here is to produce a *60 Minutes* pilot program. The entire process, from conceptual to concrete planning, is done by the students with the collaboration of the teacher. Obviously, the model outcome is seen on television. There are many ways to pick topics or themes.

The *60 Minutes* program thrives on its controversial subjects and social issues and topics ranging in scale from individual people to national and international themes. Students can suggest their own topics from their personal areas of interest within the framework of safety and practical logistics. As a teacher, you may have your own agenda for content topics, or you can give students a choice from a list of your choices. This project has a classic curricular social orientation in which to structure curricular content and process around a problem-based approach—the production of a program. A personal relevance is achieved by focusing on themes and topics within the student's own community. Many schools have video cameras and editing equipment. If they do not, numerous students also have them, and with two VCRs, crude editing can be done. After orally introducing the problem, three topics are chosen by the students. Once that is done, students are given a project description and tentative timetable for the completion of the production.

The following examples were designed to use local resources to explore how genetics and molecular biology techniques are applied in local settings. The *60 Minutes* Project and the local community make the book supplemental support material within these themes and topics.

The 60 Minutes Pilot

Topics for Consideration

1. *"Saving the Salmon: Are hatcheries no longer an answer?"*

Genetic and disease problems may limit the effectiveness of hatcheries in maintaining natural stocks. Hatchery-raised fish have problems. The hatcheries themselves may give the public a false perception of the wellness of native fish stocks. What are the real issues and the politics?

2. *"Genetic Screening: Health Insurance, Medicine, Job Security, Privacy: What can it say and what are the risks?"*

(The first problem is to get a good title.) Who uses it and what can it tell you? What is available locally?

3. *"Gene Therapy: What is the potential?"*

Is this the medicine of the future? Can disease really be cured or prevented in the womb before birth? The local teaching hospital is also a genetic research facility. What is gene therapy and how is it used?

Scenario

The class will be broken up into three groups that will become topic specialists and one small organization and leadership group. The topic specialists will assume total responsibility for the complete, unbiased investigation of a selected topic. Once the investigation is completed, teams will be responsible for composing and communicating a visual and oral interpretation of the topic in the spirit of the *60 Minutes* approach.

How?

Students will be placed randomly in teams of approximately 8 to 10 students. Students within each group will select a first and second choice regarding areas of interest within the project process. Jobs and students may overlap assignments or be modified to take advantage of the skills and knowledge of individual group members. Job descriptions may include the following:

Script writers. Presentations are scripted and logically organized to maximize visual and auditory impact. Storyboards are produced. The compositional and auditory elements of the topic must be combined in a visually interesting way.

Investigators. Resources are identified and retrieved. Specific filming sites are researched, visited, and evaluated for their potential. Interviews are planned and conducted. Individuals or groups related to the topic are identified and analyzed for their potential for inclusion in the presentation.

Production people. They will be responsible for filming and editing. They will have to work very closely with others in the group.

On-screen reporters. One or two students will assume the role of the on-film reporter conducting interviews and providing topic communication.

Organizer or leader. Each group may need only one. Leaders or organizers structure meetings and facilitate group decisions about content, production, and timetables. They will also become the spokespersons for the group in class meetings and may meet informally with other leaders and organizers or teachers.

Strategies

Teams must have a working understanding of the investigative, compositional, and communicative style of *60 Minutes*. Teams must maximize their potential to compose and produce the video production. This requires an intimate knowledge of their topic and themes and the steps, materials, and resources required for production.

Tasks

The first three days will be spent organizing the subteams, identifying tasks, and gathering information. Students will create a personal job description as a specialist within the larger group. An outline of these tasks will be used to construct a timetable for completion of group and individual work. Teams will have to maintain connections to other teams and individuals to establish a production and investigative pace. A leadership structure within the team and class will be necessary to aid organization and communication.

Outcomes

Individual Portfolios

Assessment will be based on observed class engagement in the project and the contents of a portfolio documenting all work done by the individual. Portfolios will provide evidence of mastery of content and process, overall understanding, contribution, and effort.

Individual portfolios must contain:

1. A cover letter introducing yourself and your role and documenting your personal experiences and contributions
2. A bibliography (standard format)
3. Notes
4. Copies of research documents
5. Collected material such as newspaper articles
6. Specific individual contributions to the team production
7. A self-evaluation with grade expectation

Class Outcome

The final production, combining all the individual productions, will be assessed and evaluated on presentation quality, content, depth of insight, and overall effectiveness of the communication of topic concerns.

Evaluation

Individual portfolios will be worth 70% of the total grade. The team's master plan, timeline, and general organization plan equals 20% of the grade. Your engagement or involvement is worth 10%. Your individual grade is based on the following:

> 70% portfolio: organization, completeness, depth of research, and presentation quality
> 20% teamwork: collegiality, class support, and organizational plan
> 10% involvement: rigor, depth, and degree or level of effort

The Island Project

To the Teacher

This project is heavily into island biogeography, evolution, natural selection and biological succession, physical oceanography, weather and climate, and related geology. By throwing in human activities and how they affect the physical and biological environment, you add an interesting component that could engage more students. In this form, it covers all aspects of plate tectonics. Potential locations are identified ahead of time by the teacher to limit student choice and ensure a variety of locations with the potential for authentic colonization. The locations exhibit characteristics that have the potential to produce an island and a habitat for colonizers. They are also located next to possible sources of colonizers. Teachers sometimes pick locations in proximity to islands that have established histories that can be researched by students as a source of ideas. This version of the activity also makes use of a computer software program called SIMEARTH. Some students benefit from the simulation and some do not.

Introduction

The *Island Project* is a long-term exploration of the activities and the processes that shape Earth, Earth's atmosphere, and life on Earth. By designing your own island, within the limitations and potentials of its location, you will begin to understand the complexity of Earth and its inhabitants. An evolutionary theme is a common thread that runs through the *Island Project*. Earth has changed both physically and biologically since its formation, and it will continue to change. Your island, since its beginning, has provided certain conditions necessary for living things to survive, grow, and reproduce. New land forms are slowly and continually being created and eroded. As life forms have found various ways to your island, resources were used and may have become limited. Some organisms have found ways of obtaining the resources they need. They did this through various forms of competition and adaptation. Certain populations of organisms have survived the competition and the climate; others have not. These processes are very dynamic in that the interactions between the organisms and the resources, both physical and biological, have changed over time.

Changes and disturbances to the physical environment create new challenges to many organisms. In addition, as conditions on the island change, future inhabitants and relationships may evolve further and differently. You are asked to reflect on the following biological and historical points.

Geologic Processes

SIMEARTH Assignment

Use SIMEARTH, a computer simulation program, to determine what impact present or past volcanic eruptions on your island have on the planet's atmosphere and biosphere. Be specific. List *all* of the ingredients of the atmosphere and how they are affected (look at the "air sample" window on SIMEARTH). Pull up the history window and report on changes in population, biomass, diversity, CO_2, O_2, and CH_4 as you set off new volcanoes. Next, pull up the Life Class Ratio Graph from the GRAPHS MENU and record changes in the amounts of the various life classes on your island. Which classes change first? Which are beneficial? Be specific.

The Island's Past: Birth of an Island

Describe the geologic history of your island. Based on plate tectonic theory, describe how your island was formed and develop a realistic timeline for a past, present, and future for your island. Decide on the size of your island. All islands must be at least 50 square miles (you can increase your island's size). Include rock types, age, mountains, erosionary factors, glaciers (if any), changes in sea level, rivers, and any other physical features you are interested in researching. Many islands are products of volcanic processes, but not all. Make sure you really consider the geologic characteristics of your location.

What kind of volcanoes, if any, exist on your island? Keep the nature of the local geology in mind, especially the plate boundary types. Are they currently active? If not now, when were they? Give an estimate. Describe them in detail using notes from class.

Draw a detailed topographic map and a cross-sectional view of your island. These maps should reflect, support, and aid your written discussion. Describe how the physical features (hills, valleys, rivers, etc.) have evolved over time. Use pictures, maps, or other visual aids to help answer this question. The island needs to be drawn to a scale (on graph paper) developed by you or your team.

The Island's Present: The Shaping of the Land

What erosionary processes have shaped your island? Based on your island's location, how have wind, rain, and their related processes shaped the appearance of your island and created many of its features?

The more consideration of detail, the better. Local weather (daily, weekly, or seasonal patterns) and climate really exist at your island location. An understanding of local conditions can explain and define the physical setting, and it is necessary to begin thinking about what types of living things could survive on your island.

Geologic Processes: Future

Based on your background, predict what will eventually happen to your island, using maps and diagrams. Also, describe the history of movement of the plate or plate boundary on which your island rests. Discuss the past, present, and future. Maps and charts are useful ways to depict this information.

Geologic Checkpoints—25 Points Each

1. _____ Locate your island and describe the geologic features that contributed to the formation of your island.
2. _____ Share your topo map, erosionary process information, and weather and climate data. Discuss the geologic future of your island.
3. _____ Bibliography—4 sources

Biological Processes

SIMEARTH Assignment

1. What impact would raising the MUTATION RATE and RE-PRODUCTION RATE have on the species on your island? Experiment with SIMEARTH and explain your observations.
2. What new species have adapted? Which biomes have become more common? What effect have these changes had on BIO-MASS and DIVERSITY?

The Biological Past

1. Where did your organisms come from and how did they get there? Describe how they might become successful colonizers. How did the food webs form?

2. Describe the sequence of organisms and relationships (island biogeography, biological succession, extinctions) that have existed over the geologic history of your island.

Note: You need to support your answers with specific examples, relationships, and time frames.

The Biological Present

What living things are there now? Give a detailed description of each biome found on your island. (If they are developed enough, think about food webs.) How have they evolved through time? Your living systems need to reflect local conditions. This is based on real conditions that exist at your island's location. Which plant and animal species live in each location?

Is the island in biological balance or equilibrium (ratios of sun's energy, producers, herbivores, carnivores, top carnivores, and decomposers)?

Describe your island's current food web and the ecological balance, or discuss why your island may not be at equilibrium.

Reflect on human impact. Be careful not to overemphasize it. It should be a realistic impact based on man's impact on known islands. This means if you are going to include humans, you'd better do your research.

The Biological Future

Based on a changing dynamic environment, reflect on how you see your island changing in the future. Be specific, with examples of the expected changes and the causes. Reflect once again on human impact. Give a realistic time frame for the change.

Biological Check Points—25 Points Each

1. _____ Describe your organisms, food webs, sources of colonizers, and a reasonable evolutionary timeline.
2. _____ Discuss the future of your biological systems and human impact.
3. _____ Bibliography—4 sources

Atmospheric and Oceanic Processes

SIMEARTH Assignment

1. Which of the five factors (ignore solar input) in the Atmosphere Model Window in SIMEARTH will have the greatest impact on the temperatures on your island and why? Give details of your experiments in SIMEARTH. Which two in combination have the greatest impact?
2. Which of the five factors (ignore solar input) in the Atmosphere Model Window in SIMEARTH will have the greatest impact on the biomes on your island and why? Give details of your experiments in SIMEARTH. Which two in combination have the greatest impact?

Climate and Ocean Past

How has the climate (rainfall, fog, temperature, etc.) of the past affected and shaped the physical features of your island? Physical features include shapes of mountains, valleys, location of streams, lakes.

What impact has the climate of the past had on the island ecosystem? Systems of particular interest are animal and plant biomes.

Climate and Ocean Present

Describe the climate and weather of your island. Weather includes seasons, rainfall, temperature, humidity, wind, percentage of cloud cover, and snowfall. Use graphs and charts (from Microsoft Excel).

Describe local currents, water temperatures, and waves in the ocean that may influence weather near your island. Include detailed maps of your island that contain the prevalent wind direction and magnitude across the surface of your island. Include additional maps that detail ocean currents and water temperatures.

Give details on the seasonal temperature distribution on your island. Use graphs and maps. Describe in detail the current climate of your island. How has the climate evolved through time?

Climate and Ocean Future

What impact will human interference, due to global warming, have on the climate and ecosystems of your island?

Atmospheric and Oceanic Checkpoint: 25 Points Each

1. _____ Discuss how atmospheric and oceanic processes affect the lifeforms on your island.
2. _____ Bibliography—3 sources

Communication of the Island Project

Communication of your *Island Project* information will be in both written and oral form. Your work should reflect the mastery and understanding of the natural history of your island. You are the expert. Successful projects reflect a working understanding and control of the processes that shape the physical setting and the biological interactions. You may understand it all, but you will have to successfully communicate it. To begin, prepare well for your checkpoint meetings. This allows you to focus on smaller themes without worrying about the whole project. If you do a good job on the small steps, the big project will take care of itself.

Introduction

The introduction sets the stage for your readers. It gives the reader an idea of what you are going to communicate. It also defines the points or specific areas on which you will focus. It introduces the location, current age (geological and biological), size, and any other interesting factors that would make your audience more interested in what you have to say.

Main Body

Break up the research into the major areas or questions in the project outline. Geologic time and change/evolution are the threads that tie the areas together. Major events should occur in a reasonable pattern based on time frames that we know exist on known islands. In the closing paragraph, tie events together across major research areas, such as geology and biology or climate and biology.

Conclusion

Save comments about the future of your island for the conclusion. Now may be the time to reflect on or add to information on human influence. This needs to be reasonable, however. Base your comments on human influences that you know exist on other islands.

Other Considerations

All written work will be word processed (12 pt. type) and saved to make changes easier. Figures and charts (computer produced only) need to be referred to in the body of the text. They can be embedded in the text or put at the end of your paper. Figure 1 or Chart 1 should be referred to first and placed in the back of your paper in numerical order.

The bibliography needs to be complete, including page numbers. There are formal bibliography formats to follow.

Evaluation Criteria

This assignment is designed to allow students to research and apply the knowledge from class and individual research. It also will help you learn and model skills in structuring a long-term project. The length of the project requires planning and group and individual responsibility. The individual questions are included to help structure your time and focus. Long-term planning and structuring of your work time will reward you with points.

Checkpoints 200 pts.
Rough Draft 100 pts.
Final Draft 100 pts.

What Can We Learn From the Past?
(or Lessons From the Dinosaurs)

Introduction

The estimated age of our Earth is 4.5 to 5 billion years old. Larger life forms have existed only for .5 million years or so. The age of dinosaurs covers roughly four periods of geologic time. The four periods, from early to later, are the Permian, Triassic, Jurassic, and Cretaceous. The dinosaurs' time frame runs from 70 to 230 million years ago, roughly 160 million total years. If we measure species success by longevity, then dinosaurs must rank as the number one success story in the history of land life. In contrast, mammals have been the dominant animal on Earth for only 70 million years and humanity, in one form or another, only 100,000 years. The exact and relative dates are somewhat blurred and vary from book to book. People today are watching the extinction of species happen at what seems like a rapid rate. At the same time, the human population is increasing. This leads many people to wonder if our own environment and the human species can sustain such growth.

In an effort to understand possible future events, we need to understand past events. The dinosaurs became extinct over a 1- to 2-million-year period—a relatively short time—and the reasons are being researched and explored. Our major goal is to compare and contrast the events and conditions during the targeted time periods and today. After completing this, we may be able to reflect upon and predict future life and environmental trends.

To accomplish this, we are going to send our best student researchers back in time to get a clear picture of the events and conditions during the target periods. We will divide our teams into four major groups, each researching a period. We will add a fifth period, our own period: the Quaternary. This will provide us with a basis of comparison with earlier periods. Within each period, researchers will specialize in the following areas: group leader; animal specialists; plant specialists; ecologist; atmosphere, climate, and water specialists; and geochemistry/timeline specialist. Each group will work together to complete a big picture of its time frame. The outcome will be a written and oral report describing the time frame.

Job Descriptions

Group Leader

The major function of the group leader will be to coordinate group activities and write or combine other group members' work into the final report and coordinate the efforts of the group. The group leader should be able to keep the "big picture" goals of the group in mind. Remember, all of this research will help us make predictions for the future of the Earth and humanity, which will be the final product of this research. He or she must be able to facilitate the group becoming an expert on its time period so that the class receives a clear picture during the oral report. He or she will also need to help group members as necessary, keep track of the location of group files, and help find references. It also may be necessary for the group leader to fill in group gaps. A gap may be work left by an absent or less motivated group member.

Evaluation will consist of weekly progress report "check-offs" and the final report and paper.

Animal Experts

Each geologic period covers 40 to 75 million years. During each period, animals and their relationships change over evolutionary time. The animal experts will document all the animals, not just dinosaurs; their relationships (with other living things); and changes that occur (extinctions, new species, or changes in behavior and appearance). Document the numbers of species from each animal group and include a general description of the animals' lifestyles. Relate them from an evolutionary standpoint with animal groups from the periods before and after theirs. Was the rate of extinction equal to the development of new species?

Plant Experts

Like the animal experts, the plant experts are looking for the same understanding of the plant kingdom.

Ecologist

The ecologist looks for connections between living things. Define the major relationships between plants, animals, and their physical environment. This will require collaboration with the other specialists. Describe the energy flows between organisms, food webs, and chains. What were the migration patterns and the reasons for the migrations? How did different species handle seasonal changes? Look at all the current explanations for the extinction of the dinosaurs and changes in the environments that favored mammals.

Atmosphere, Climate, and Water Expert

Understanding the physical environment and patterns of change in that environment is necessary to understand and define the individual species stress factors. Specific items that need to be included are worldwide temperature patterns and changes, changes in sea level, and rain and drought patterns during the target period. Reflection about impacts and connections must be backed up with data.

Geochemistry, Geology, and Timeline Specialist

To understand the changes to the environment and the living things in it, we need to put major events in a time context. (What happened when?) Geological activities are an important part of understanding the conditions of the time. This specialist will work with others in the group to construct a timeline of major events during their period. Also, he or she will need to work with other group specialists so that no holes are left in the overall timeline. Another area of research will be to help the groups understand how scientific research is conducted to answer questions about the period. A collection of questions, research, and answers will be compiled for examples on how humanity comes to an understanding of past events.

The Process

Organizational skills can help you identify and work through tasks necessary to get through large assignments. Making productive use of your available time and breaking down tasks into manageable subunits works to reduce those feelings of being overwhelmed by large amounts of work. The following timetable will help you begin to organize this project. Each week, you have the opportunity to earn points by completing the described tasks. Please feel free to do more than just those required tasks. These tasks are just to help guide and focus you on short-term goals that will lead to a solid background about your subject.

Each group will meet with the classroom teacher once a week to present work in progress. These sessions allow you to exchange ideas and receive feedback from the teacher. Points will be awarded based on evidence of completing the following intermediate tasks. The total project will be worth 200 points. Out of the 200 points, 75 points (25 per week) will be given for *evidence* (notes, bibliography cards, rough drafts of reports, etc.) of progress. A rough draft is due at the end of the fourth week and is worth 25 points if the rough draft requirements are completed.

Week 1

Group Leader

Word processed up-to-date bibliography of the group's references.

Specialists

Project specialist work outline, which begins to organize your plan and information for the written and oral report.

Four reference sources should be documented on index cards. Specialists must have these on record with the group leader to get credit.

Week 2

Group Leader

The group leader should have notes documenting group and individual progress and a list of individual goals for the next week. A continued up-to-date bibliography is needed.

Specialists

Specialists should present the evidence of research progress on specialty questions that demonstrates a clear understanding of the questions and the beginnings of their research.

Week 3

Group Leader

A statement is required of the group's overall progress and an outline of remaining tasks. An outline for the group's oral presentation is also due at this time.

Specialists

The final bibliography and written introduction are due. The introduction needs to stress what you are attempting to have readers know and understand about your topic.

Week 4

Group Leader

The final oral presentation outline is due, as well as the final statement of progress. The group leader should have hard copies of all rough drafts to help begin the critique and identify weaknesses in overall research. Group leaders need to have access to computer files at this time.

Specialists

Rough drafts are due, with a copy given to the group leader. Let group leaders know the locations of your files.

Photosynthetic Green People Project

To the Teacher

This scenario is designed to provide students with a purpose for learning the concepts of photosynthesis. It is a good introduction to problem-based learning. Outcome expectations can be tailored to the philosophy of the teacher and range from the "How does this work?" details with associated vocabulary to more of a general introduction. It is also a good way to introduce the physics of light and how it relates to biological processes. There are many pathways—some integrate concepts that could be developed within the project.

Math components can be designed to look at carbohydrate produce, potential calories, and skin surface areas versus body volume. We have even looked at an individual plant surface area versus its volume and compared it to a human's. You could even go into how much surface area would need to be exposed to provide any given amount of carbohydrates. It is all fictional but makes a good intellectual puzzle.

Scenario

There was a laboratory accident several years ago in a research facility known for its plant research. Exact details are not clear, but we know that several male and female graduate students were injured. At the time, they were amplifying various amounts of DNA from a number of different plants. The DNA was to be used in recombinant DNA plant projects. The general idea was to produce a plant with more efficient photosynthetic structures. This has the potential to increase growth rates and plant productivity.

A few of the graduate students went on to marry and have children. Here is the interesting part. A few of the children were found to have skin tissue containing chloroplasts. Each child and his or her parents had their genomic DNA screened, and it was found that their genomic DNA contained genes thought to be found only in plants. The possible mutation and/or inserted genes were found in the parents' and children's germ (gametes) cells. Scientists are now very interested in how this occurred and what it means for the parents, children, and their future generations.

What?

The local paper's science editor has asked this class to write a series of newspaper articles that describes what has happened and how the biological mechanisms for the changes work. The editor has broken down the articles into various subtopics in which she feels readers might be interested. They are described as follows.

Biotechnology—An Accident Creates Some Interesting Questions

In this article, the accident is reviewed, and the mechanisms for such a phenomenon are explained. How could this happen? Most of the public does not have the knowledge or background to understand. This is to be the public's introduction to biotechnology, genetics, and molecular biology. The exact details are still secret, but with sufficient background, the writers can form several scenarios in which to explain what might have happened. This is not a fictional article. All ideas need to be supported by what we know about the science behind it. Explain how their photosynthetic abilities might work. The structure and function of light, skin, chloroplasts, and photosynthesis need to be tied together.

Biotechnology—What Is Happening Today?

In an effort to keep the readers informed, this article should reflect on current applications of molecular biology and genetics, and how they affect our everyday lives. The writers may want to include other examples of applied recombinant DNA projects. How is this technology being used today? Some people have raised concerns over the ethical and moral issues involved in biotechnology. What are the issues? Are there risks involved? Certain people perceive dangerous consequences for this technology. What are the true risks of this technology?

Biotechnology—The New Green Photosynthetic People

Now that the public has a sufficient background, what should happen to the people involved? Should they be allowed to reproduce? What might be the benefits of this accident? How might it change their lives? Are there negative consequences? If we were able to continue and improve the technology that created and altered these people, how might we go about it? How could we maximize the potential for any benefits?

Biotechnology—A Green Photosynthetic Society

This article is to reflect on the issues that the green people present. How will society deal with this new group of people? Will they be accepted? How could these traits become more prevalent in human populations? If these traits are beneficial, how might they change human behavior? Where would these people want to live? How could this change an individual's diet? All ideas need to be supported by current actual scientific knowledge.

Organization, Leadership, and Communication

All the stylistic expectations, editing, standardization, and ongoing project decisions need to be addressed within this group and communicated to the others. This group will take care of all the details and create and provide a coherent structure for other groups. Individuals may be asked to aid groups with specific problems and fill in when necessary. The leader will collaborate with other groups to develop a working timetable and intermediate points of accountability. The group will work with the instructor to establish goals and objectives, and it will be responsible for the final presentation.

Tasks

The first few days will be spent organizing group and individual goals and gathering information. Students will need to break down tasks into identifiable categories or subgroups. Once these categories are identified, students will be free to pick specific tasks for research. Some tasks may require more than one student. The first tasks for subgroups will be to brainstorm, design, and develop strategies, goals, and outcomes for their areas and some justification for their outcome's contribution to the project. An outline of these tasks will be used to construct a timetable for completion of group and individual work. Specialty areas will have to maintain team connections to support the total group effort.

The leadership structure will be responsible for organization and communication during this period.

Research Strategies

The questions are created to help student groups focus on useful concepts and information needed to fully master the ideas and mechanisms within the project. The background gained from the answers to these questions enables students and teachers to exchange and apply information and to clear up misinterpretations, and it allows teachers to make suggestions for further work. Teachers also act as research guides and aid in assessment of the mastery of concepts within the project.

Past experience has shown a crunch of students at the end of the week with few discussions earlier. Consider this when scheduling your study time. If you wait until the last minute, you may not get in, hence no points.

Questions for Guidance

Each group should consider at least four questions per week that more directly apply to its focus. The group should discuss these with the teacher and have them checked off. Each question is worth 10 points.

Human Ecology

1. How would it change the average human lifestyle and where we or they choose to live and why?
2. How would these characteristics become more or less prevalent in the general population? Be specific about the mechanisms that would change the numbers. Use real numbers to explain your answer.
3. Would these traits be selected against or for? In all parts of the world? How might cultural selection play a part in this new subspecies? How could disease play a role in the development of this "new" human?
4. Project the outcome of adding 100, 1,000, or 100,000 of these people to the worldwide human gene pool.
5. What could the changes mean to our current use of natural resources, agriculture, and economy?

Biochemistry

1. What are the products of photosynthesis and the amounts produced?
2. What important molecules (chemicals) could be produced (by photosynthesis) that humans could use in their own metabolism?
3. As you know, the main groups of organic compounds important in living things are carbohydrates, lipids, proteins, and organic acids. Which of these groups will increase in importance if we switch our energy source from ingestion of food to photosynthesis? Give a detailed explanation of why. Which of these groups will decrease in importance? Why?
4. Draw line or dot structures of the 30 most important organic molecules to your Photosynthetic Human. Detail what functions they perform (i.e., metabolism, energy transport, structural, energy storage, etc.) and where each is located.

5. What other important elements are needed for photosynthesis to take place, and where could a person get them in his or her diet? How much would we have to supplement?

Challenge Question

What are the chemical products of digestion and how would their needs change?

Math

1. Research at least two different types of plants and determine the photosynthetic surface area in relationship to the total volume of the plants.
2. How much area and at what density would the chloroplasts have to be concentrated to provide all of the dietary requirements of an individual? Do we have enough surface area to meet our total dietary needs?
3. How much oxygen or starch does a plant produce per any given photosynthetic area? Compare that to the area of humans covered with chloroplasts.
4. What are the rates of photosynthesis necessary to provide desirable products? How long would you have to expose your chloroplasts?

Challenge Questions

1. Would a less direct (plugging in to plants at night) symbiotic relationship with plants be more efficient? For starch? For oxygen? Compare this idea with the photosynthetic human idea.
2. How would the rate of photosynthesis vary from the equator to the poles?

Individual

1. How might internal human physiology and anatomy change as a result of less need for certain biological organ system function?
2. How could this idea change the way humans might structure our days and nights? Consider both work and play activities.

Team members must cooperate to identify and obtain resources in a timely manner. Documentation of sources is essential. A timetable may be necessary to keep the work on schedule. Intermediate outcomes must be identified and are required by instructors in addition to the final product. Sharing and cooperation can contribute to your overall success.

The four large groups and the smaller leadership group will submit a final product. Each group's product, with each contributor's specialty outcomes, information, or documentation, will be placed in newspaper article form and become part of a package sent to the newspaper.

Individual Student Outcomes

Portfolios

1. Each student will maintain an individual portfolio documenting all work done for the articles. This applies to smaller subgroups as well. We want to see your individual contribution.
2. Each individual or subgroup will complete its own file of information supporting the group's work. It will be turned in and assessed at the end of the project.
3. All teams must agree on a common style guide so that fonts, type size, format, and organization of the final document are consistent.
4. Individuals will provide evidence of mastery of content and process, overall understanding, contribution, and effort with their completed portfolios. This, in addition to intermediate outcomes, will contribute to individual assessment.
5. In addition to the article outcomes, all students will be required to take a knowledge-based test to demonstrate mastery and understanding of the concepts explored in this project. Be prepared!
6. Individual portfolios must contain

   a. A cover letter introducing yourself and your role and documenting your personal experiences and contributions
   b. A bibliography (standard format)
   c. Notes

    d. Copies of research documents
    e. Collected material such as newspaper articles, travel brochures, and so on
    f. Specific individual contributions to the team rally manual
    g. Copies of intermediate outcomes
    h. A self-evaluation with grade expectation

Assessment

Individual portfolios will be worth 80% of the total grade. The team's manual equals 20% of the grade. Your individual portfolio grade will be based on the following:

    70% portfolio: organization, completeness, depth of research, and presentation quality
    20% teamwork: collegiality and team support
    10% involvement: rigor, depth, and degree or level of effort

5

Doing Authentic Science

Introduction to Mentored Relationships:
Taking the Classroom Into the Real World:
One School's Approach

◢ The program and experiences described here were the outcome of the development of a successful interactive community program within a smaller academic "school within a school" at a high school in northern California. The high school, of roughly 1,600 "salt of the earth" students, is divided into six smaller, self-contained learning communities that offer various educational experiences and pedagogy. The school is considered a "restructuring school," and it is affiliated and associated with various restructuring organizations, programs, and models. The school actively seeks and has benefited from grants and other forms of support. Students, with their parents, are able to select the smaller community that most nearly meets their personal educational goals and needs.

The smaller school community of about 250 students and 8 teachers is called C-TEC, The Center for Technology, Environment and Communication, and it uses an active mentor program in the sciences. The educational areas or specialties that define this smaller school community include an applied use of technology and our ability to embed more traditional content and process into classroom activities that are inherently more interesting and motivating than more traditional pedagogy. Teachers in C-TEC often use environmental concerns as a focus or vehicle for activities because of that topic's ability to engage students. Emphasis is placed on the skills and techniques that clearly and effectively communicate ideas, thoughts, and results in real-world, authentic applications. Students gain a new appreciation for their work because outcomes are shared and critiqued by a wider audience. For example:

- A video production on nonpoint pollution and its effects on the biologcal community was created by students and shown on a local television station.
- A student project group monitoring groundwater levels and other issues relating to gravel mining on a local river presented their research findings to the local County Board of Supervisors. This included the effects on the river habitat.
- Pamphlets, created and written by students, were included in local water bills alerting citizens to water saving and pollution problems.

In addition to the above-mentioned areas, students are most noted for their community mentor project work. This is the most visible part of C-TEC's program and the area that receives the greatest attention and interest. This is the part of the program that is defined here.

There are currently about 180 students in year-long project classes. These classes are the main connection to the community and mentors. These classes maximize the potential for learning in off-campus settings and using professionals in many areas. In addition, students can be engaged in working on problems that could not be brought into the classrooms.

For C-TEC students, a project class is a class with a teacher that offers a variety of year-long educational paths for discoveries and explorations within that teacher's expertise. Offerings vary from working at a marine lab or wastewater treatment plant to video production, computer animation, or criminal law. Each year, there are

10 to 15 different project areas from which to choose. Some students find their own mentors and projects. Teachers act as a major resource or coach in creating, planning, implementing, and communicating their project work. A richness and relevancy is added as teachers and their students also work with community professionals who mentor students in their projects.

Mentors bring to the school a variety of unique educational opportunities that allow students to work in specialized areas on authentic problems that the mentors bring or develop with the students. The mentors connect school activities to the real world outside the classroom. These experiences are driven and motivated by the individual student's or student group's own special interest. Students use a portfolio structure to organize the design, development, implementation, management, and communication of their experience. The structure of the portfolio is described in later sections.

There are many steps to the creation of a successful mentored project. Planning is a collaborative effort, and now more than ever, it includes parents and administrators. Projects are open-ended and usually offer something for all ability, interest, and motivation levels. Each project may offer opportunities for an individual, a few students, or many students. Usually, there is a variety of opportunities within each area. The following are samples of the project topics that students are offered:

- There is a working relationship at a University of California marine laboratory that has researcher/teacher mentors working with students on many life science topics.
- A city engineer included C-TEC in an Environmental Protection Agency grant. The grant helped create a nonpoint pollution test site at Piner High School. A public education component allowed students to design and create a TV commercial and water bill inserts alerting the public to nonpoint pollution problems.
- A chemist from wastewater treatment is working with students on sewage.
- Video production and animation teams are creating original productions and documenting research projects.

Motivation to include mentors or community resources can come from a variety of sources. For example, a biology teacher would love to connect his or her enthusiastic biology students to a working laboratory in a local college. A local hatchery is also a research site for a state fish and game agency. Biologists there are using molecular techniques to investigate a genetic bottleneck within a species of salmon, and a number of students have asked if they could work there in some capacity. These are both examples of motivational education activities that have not fit the traditional classroom paradigm but could fit into a mentor program.

Mentors in this school's context are usually described as working science professionals from the local community who are willing to work with students. In the past, outside science professionals limited their involvement in local schools to career presentations or one-day lectures on topics within their expertise. Sometimes, student visitations, field trips, or job shadowing arrangements were made to allow students to visit off-campus sites. The idea was to have science professionals expose students to the inside workings of their jobs, businesses, companies, and organizations. They interpreted what they did and the preparation it took to obtain various positions within the profession. Other experiences included presentations on specific topics such as health, medical, or other science topics. Teachers began to see these science professionals as an underused educational resource. They saw a potential to create relationships that could be much more rewarding for the teachers, mentors, volunteers, and students. Mentors could bring their real-world science problems into the classroom to challenge students. This is the essence of the problem-based approach.

But the mentor program fills other needs as well. There is the basic premise in public education that the things you learn in the school setting will be necessary in later life. The relationship between what goes on in most classrooms and the skills and knowledge required for success outside the classroom is based on the faith of the students and the parents that the curriculum is valid. Rarely are the connections clearly visible for students. How many times have you heard students ask, "Why do we have to learn this?" In education, acquiring knowledge and techniques and understanding processes is important, but applying these qualities in an authentic context ultimately determines and defines success or failure in most real-world settings.

Most science professionals are problem solvers. That is what they were hired to do. Lawyers, scientists, engineers, writers, and many other professionals are presented with problems that require past knowledge and skills in addition to new knowledge and skills as old methods become obsolete. They continue to learn as a matter of necessity. This is especially true in science. The thread of the problem-based approach can weave through a mentored science relationship naturally. In this model, these professionals bring problems with them for students to work on and solve with the support and collaboration of the mentors and teachers.

Program designers wanted to engage the mentors, yet do so in a way that would not intrude into their lives or detract them from their jobs. But they also wanted to foster a more long-term, personal teacher/mentor/student relationship. We found that they needed to educate science mentors in new educational strategies as well. The principles of problem-based learning (student-centered, self-directed, open-ended, etc.) had to be built into the mentored relationship structure. Mentors brought authentic relevance and context to the learning experience. They also began, many times, to provide resources that are unavailable in most school settings.

The opportunity to begin mentor programs exists in every community and can become an important part of any secondary school experience. Teachers want students to have these experiences in a safe and supportive environment. The goals or outcomes of reading about this community program are to open the doors to community science educational opportunities and partnerships for teachers, mentors, and students. The following examples and experiences can be used to help you design and adopt a new program in your community.

A section specifically for potential science mentors has been included. If you find it useful, please feel free to duplicate it and use it as an introduction to the mentor partnership program. You may also need to modify it to reflect your needs and vision for the collaborations and partnerships. In addition, samples of current science project descriptions and student insurance waivers are included.

The potential for many self-motivating, purposeful, enriching, and relevant educational experiences exists in every community, not just in the classroom. Creative teachers can make the connections and make these experiences happen.

The community has provided this school with many successful mentored scientific educational opportunities and experiences. The intention here is not to expect all teachers to do what C-TEC does; there is no set formula. Teachers, mentors, and students need to know what is possible and build on it. For some teachers and schools, the program described here can be meaningful and relevant and seed the creation of a mentor program that will provide a unique richness.

Outcomes for Mentored Relationships

One of the basics for planning long-term teacher, mentor, and student experiences is setting expected outcomes for the collaborations and interaction. For students, outcomes could include gaining the knowledge, techniques, and processes necessary to answer a scientific research question through investigation and then writing a paper, completing and applying a computer program, writing and publishing an informational coaching and sports medicine manual, and documenting a research project with video or presenting the results of research to the authentic audience. These are important outcomes for the students, but teacher and mentor educational outcomes are very different.

Orchestrating self-discovery, exploration, and a rewarding educational experience are our outcomes. This program's agendas are hidden. Content and process can be carefully embedded in the project experience. Frustrations can be placed selectively and avenues for success created. Student, teacher, and mentor project collaborations should offer experiences rich in content but also rich in opportunities to learn how to get work done in an organized, effective way. In addition, they should offer paths to experience and overcome problems inherent in any project typically experienced in most work sites. This would include modeling solutions to the frustrations we all face at the workplace and helping students build the skills and confidence to deal with and work through these frustrations on the road to a successful project. Mentors and teachers are not just telling them what they need to know. They are building both a customized program and a structured path within the project that are motivating and interesting and then coaching the students on their journey.

Recruiting Mentors

Mentor recruitment is an area that inhibits many teachers who are otherwise interested in doing project work. From our experience, mentors are easy to find. Most people want to help but do not have a clear vision of how to do so without intruding too far into their time and responsibilities. Careful consideration, thought, and planning can alleviate their concern. We usually have more mentor opportunities than we have time to pursue. The key is flexibility and a willingness to work with what the potential mentors are willing to provide.

As a science teacher within C-TEC, I look into my community to see where science is being done. I also look into my curriculum to see where mentoring could fit and enhance learning and teaching. Many times, I lack the resources to cover some topics as well as I would like, and a mentor or community source may help. Sometimes, there is just something interesting going on that I want to include, and so I target that.

Various scientific businesses and public agencies exist in most cities and towns. Science is being done in most community colleges and universities and in governmental agencies such as those concerned with water and air quality, fish and game, and forestry. For those in other disciplines, law is being practiced in every aspect of community life. Papers are published and products are produced. Research is being done in most college and university disciplines. I had two seniors working as urban archaeologists who were going to a local college and sorting through soils collected from an Oakland, California dump in a historic part of town. As part of their work, they were to document the evolution of this study from funding to local politics in addition to the science. There are also museums, courthouses, and public works departments with engineers and city planners. These are all good places, both public and private, to look for mentor project activities.

Businesses and public agencies will contribute professionals in the form of donated hours. Sometimes, they will allow the use of technologies and equipment that schools cannot provide. We do not overlook our parents, either. As we describe the program during parent nights, interested parents are asked to sign up if they feel they have something to offer. Parents are a primary source of contacts.

We approach all businesses and public agencies the same way. We do not have a set plan for involvement, and we do not ask for anything. Once we communicate the nature of the program and provide a range of examples, both sides begin to see the possibilities. Most organizations are willing to work with us; we want their help and cooperation on their terms. Mentors usually end up putting in much more time than they originally felt they could. The process of recruitment is outlined here; it does not always go like this, but this has been the pattern after a couple of years of recruitment.

The First Visit

Our project program, with examples, is presented during the first visit to a potential project mentor organization or individual. The first person you talk to, in most organizations, may not end up as the mentor. After the initial visit and contact, the person you talked to about project work usually shares it with others in the organization or delegates further discussion to people who might have a greater interest in participation in your program.

The Second Visit

Usually, this visit brings in the potential mentors. They are generally afraid you are going to ask for more than they are willing to provide. Your job is to be "user friendly," take what they are willing to give or do, and see if it will work for you. We again define the nature of the program, clearly and completely. This includes time commitments and a discussion about what other mentors have done in the past or your personal vision based on your needs. Many people think you want them to come in and present their work or set up job shadow or visitations. Be clear about your vision. Working with students is very different then a simple visitation or presentation. If the discussions go well, they can evolve into more concrete planning.

Collaborative Project Design

We then set up another conversation after both sides have had time to reflect on the possibilities. Invite them to your school, share ideas and new possibilities, and discuss limitations. We want mentors to help orchestrate paths of self-discovery within real problems from the students' own community and the mentors' own workplace. Sometimes, it will not work, and that is okay. The chemistry is not always there. Experience gives you a better feel for the potential of a relationship. If things go well to this point, the relationship can become more concrete, and real planning and collaboration can occur. Now is the time to decide on the details. Start with expected outcomes, such as what you want your students to know and do at the end of the project. Share these ideas with the mentor and then create the paths to get there.

The number of mentors with whom you work is up to you. It may take a few to meet your needs. When and where they mentor is up to you. Some project activities occur off-site, whereas others occur at school. It is nice to have options so that students without transportation can participate.

Liability and Insurance Issues

Insurance and liability problems can be worked out with your school administration and the mentor agency. Additional district liability is scary for administrators, but there are models from athletics and work experience-type programs that can be examined for precedent. Your district business manager is a good place to start. In our case, our principal set up a meeting with the business manager and district insurance carrier.

Once project activities and student participation were clear, we shared it all with the appropriate district personnel and insurance carrier. We described the activities and eliminated or changed those with which the district was not comfortable. In one case, a mentor wanted students to take samples of a specific species of fish for a population census within a lake, which would require students to go out in boats. We did not do that project, but rocky ocean intertidal research was fine with appropriate preparation. A well-planned presentation, a good modeling organization, and control over the activities goes a long way to put district administrators at ease. This approach also works with parents.

If students are to work at facilities and sites with mentors outside the school, each mentor's organization may have its own insurance protocols. We had students working at a University of California laboratory and field study site. The university required its own release and waiver forms signed by students and parents.

After these discussions, we design necessary waivers and release forms. Students participating in sports or work experience programs can provide a model for off-campus activities. See how teachers handling those programs deal with these issues. Be prepared with answers to administrative, mentor or mentor agency, or parent concerns before they ask. Plan parent meetings with mentors present. Identify the risks up front, work through them, and mitigate them or eliminate them. Many administrators will see potential trouble with many activities unless you show them you are on top of the potential problems.

In some cases, you may have a student or parent who is just not comfortable with an activity. You will have to have options for them. In our school, off-campus projects are only an option, and we have to have on-campus alternatives. Laboratory work, maintenance of collected plants and animals, computer programming, multimedia, literary production, and many other project-related activities can be done at school. In some cases, half a group may work off-campus and half on-campus.

In closing, flexibility is the key. Project opportunities usually end up to be a compromise between your needs or vision and what the mentor or his or her agency will provide. There is no set formula for recruitment of mentors or the creation of a project except for keeping an open mind for opportunities for your students. Working with individuals who use what you teach in their jobs keeps you fresh, current, and aware of the skills and abilities that are required in the real world. Sometimes, we forget what we, as students, thought was important and motivating in our school experience. When you think back to when you were in school and think about those few memorable activities, what really was important to you? Rarely was it a chapter in a book. Mentors and mentored projects can stand out from the day-to-day curriculum and make a real difference in a student's life and how he or she views the community.

To the Mentors

I have changed my writing style to address a new target audience. The focus moves from teachers to mentors in the following two sections. I am writing to the potential mentors in the style and language that I have used in recruitment of mentors. It may be confusing at first but it does capture the essence of the recruitment process.

We want your help and invite you to explore, with teachers, potential ways to provide educational experiences to our students beyond the classroom. We are offering you the opportunity and structure to share what you do with students interested in your field. When the community becomes the school, everyone wins. What if the community got behind academic opportunities the way it gets behind community and school sports? Usually, whole communities get behind extracurricular activities. We all know learning does not begin or end at the classroom door. Most of us learned what we really needed to know once we got into our jobs. Schools gave us the basics and sometimes the interest and motivation to continue into an occupation of our choice. If we were lucky, we had a few people in our lives who believed in us and helped create and support our dreams. Anyone can fill that role, not just teachers.

Acting as a mentor or providing a project has the potential to have a major positive impact on our students. Professionals know, usually better than our teachers, what it takes to be a scientist, writer, computer programmer, or video producer. Many businesses and public agencies play a role in shaping and defining the identity of a community by sponsoring sports teams and participating in civic organizations. A community is defined by the quality and caring of its residents.

We, as teachers, want to bring the community into the schools in new ways. Mentors and long-term connections and relationships during collaborative projects unify our community's educational efforts. We want mentors to know what we do, know our students, and know our problems. Together, we create avenues to solutions. We cannot teach unmotivated and disinterested students. Mentors and long-term projects connect students to the realities of the workplace and the knowledge and skills they require. You make our curriculum relevant and alive. Collaborations with mentors create a more authentic environment for experimentation, exploration, and learning that beats a "chapter march" and worksheets every time.

There are many levels on which community agencies, businesses, and professionals can contribute. You can contribute to local schools in ways that will be rewarding for you, that will not interfere with your business or job activities, and in some cases, may provide you with tangible benefits. We hope you will consider participation and collaboration with teachers, which can begin to inspire creative possibilities that can work for everyone—most importantly, the students.

What We're Looking for— Project Methodologies for Mentors

We want you to help us build opportunities for students to apply what they have learned and experienced in the classroom. Creating a safe and structured environment supports a high potential for a successful experience. This section should give you an idea of how we do it. We recognize that a mentor program can be an intricate part of a student's classroom or school experience. We also recognize that many mentors, while experts in their own fields, could be uneasy about working with teachers and students. Mentored relationships are very unique and may include curriculum and teaching methods that you may not have experienced. We hope that we can give you a brief idea or recipe for overcoming any reluctance or concerns you have. We want to help structure and implement the project with the greatest chance for student success and provide a rewarding experience for you.

Things you can count on:

1. Teacher support in managing and structuring the project, student accountability, parental involvement, evaluation, and final communication of the work

2. Collaboration on your timetable. There is no minimum or maximum time commitment. Your relationship and contribution is open-ended. We have mentors that communicate with students via e-mail or telephone only, and we have mentors that visit the school or have students come to their work sites. The nature of the project interaction and communication is very flexible.

3. Parental support, at least for transportation, and sometimes more help
4. Open-ended opportunities in structuring your project on and off campus

The idea, in this type of educational experience, is to foster a collaborative relationship where students take some responsibility for choosing an area that interests them, and the mentor helps by selecting problems, experiences, and projects from the real world. The collaboration then plans, structures, implements, modifies, and communicates the outcomes of the work. Roles for teachers and mentors are defined. The work can be original, or you may know the outcomes, but the students do not need to know that. Teachers want the students to feel their work is important to you, and we want them to take "ownership" and have it become theirs. Ideally, we are orchestrating and building paths of exploration and self-discovery and adding a little mystery for motivation. We want to avoid the "canned labs," lectures, and worksheets and instead model how work gets done in your workplace. If we construct our activities well enough, traditional content and process will be needed to solve a project step or problem; thus, the project activity fosters and motivates the students' need to know. Then, at this point, we teach and coach. Knowledge and technique become important tools to solve interesting and motivating problems. It becomes much easier to teach when students want to know and want to apply what they know.

In addition to this, we also want to integrate and connect the curriculum from all the major school disciplines. We recognize that most things taught in isolation really work together everywhere else but in school. Please have high expectations for student communication, math, and other skills required in the project. They may need to be taught the proper protocols appropriate to your area. Also, feel free to help students understand some of the moral or ethical questions that may apply to their project.

Planning sessions with all involved—students, parents, mentors, and teachers—help all participants to come to common understandings about the nature of the experience. This is an opportunity to define your expectations, identify risks, or cover other areas of concern. We try to limit the misunderstandings and problems before they begin. Parents do not remember school being like this. It is new to them. Teachers will explain to them that we want to simulate real job site skills, collaboration, cooperation, deadlines, resourcefulness, and so on. This helps them begin to understand our major goals. Students are not used to taking so much responsibility for their own learning, and parents no longer just drop their students off at school. It is a new way of learning and teaching in an old system, and we need them to buy in. They have to be taught to learn like this. Most of us learn on the job and rarely return to a classroom setting after completing school. The way we, as professionals, learn on the job is similar to what we want our students to experience.

The Teacher's Role
Within Mentored Relationships

To begin, mentored project work takes more work than the organized "chapter march." Project teachers have to be creative, open to people that may know more about their subject then they do, and able to think on their feet as unexpected problems come up. Your role changes from a provider of information to providing structure, support, and connections to the resources needed by the mentor and students. You create the vision for the collaboration and the project. The rewards are great. Students will take ownership of the project. You may no longer be a "stand-and-deliver" teacher. The mentor becomes the content expert. You now have the responsibility to create the paths of exploration that the students and mentors will take. Mentors generally do not have an educational background. They are successful at what they do and want to share their experience with your students, but they do not always know how. This is where your expertise comes in. They will need your help in designing outcomes, creating timetables, evaluating work, creating accountability, and setting realistic expectations for your students. You know your students—place them well. They need paths built and goals created that are reasonable and attainable. Not all students come with the same toolbox. If they are to work in groups, consider the group dynamics and chemistry. Project topics need to be chosen for their richness, real-world relevancy, and their ability to motivate and interest students with a variety of abilities and needs. This is a collaborative effort, but you are the expert.

Another benefit that more directly affects the teacher is the teacher/ mentor interactions. As a teacher with mentors, I look at the time I spend with mentors as inservice time. How many inservices, staff meetings, and committees have you been involved with that allow you to work with people on the cutting edge of your discipline? Teachers rarely have opportunities to interact with colleagues in other schools. We generally work in a very narrow and closed environment. Teachers work with other teachers and rarely or never are given the opportunity to work with people that really need to apply what we teach. Interaction with mentors keeps our perspective fresh and in touch with the real and applied state of our discipline.

Project Planning

Planning for projects, a problem- and theme-based pedagogy, is an ongoing activity for us. As with all problem-based activities, education is seen as a process (teaching students how to learn) as well as a body of knowledge. Most teachers do have mentors work with classes, but it is more typical to have mentors working with their students in their projects class. The projects class is an elective that allows students to work on problems or programs of their choice, usually with a mentor and sometimes off-campus. They may work at a local lab, field study site, or other community location based on the nature of their project focus. This is something like a work experience model, except it is more academically focused. The class itself gives a student credit in a discipline that most clearly fits the student's project activity and level of complexity. The closest model that fits or looks like our project model is an independent study or special project class in a college or university setting.

The projects class works for C-TEC, but I do not envision a projects class in every school. However, the project idea could become a big part of a more traditional class. Usually, the first argument or concern that comes up from teachers is "I already have too much to teach," and I agree. Even when I taught in a more traditional way, I could not cover all the curriculum. I covered what I was more successful at teaching or knew best. Giving up chapters became easier as I worked to embed the content in the project experience. A carefully constructed project will drive the need to know the content. That allows me to cover those lessons that did not lend themselves to the project's context.

Again, this is where teachers new to this teaching style have trouble. They feel, and sometimes the parents feel, that students miss out on important material. It is my feeling that students remember and retain a curriculum in which they are sincerely engaged. What did you remember from your high school experience? What activities prepared you best for college or your profession? Our goals are knowledge, content, and sometimes process, but the most important goal we have is to create enthusiasm, motivation, and interest. These intangibles will carry students further than hitting every chapter in the book. I do not advocate doing away with more traditional teaching, but I think we need to look at what is really valuable and important to students' future success and find a balance. Is it sometimes harder to teach this way? Yes, it is, but it is much more rewarding.

The Project Portfolio

The entire year-long project experience is structured and held together using an instrument or organizational structure called a portfolio. It is a device that gives the project work structure, form, and built-in accountability. The portfolio itself may be just a file folder filled with information about a student's project work, or it could be more. Each teacher will have his or her own requirements. The portfolio is the way a student communicates to mentors and teachers. It documents the process of doing successful project work. Most professional people are good at structuring their work duties and become very efficient at getting the most out of their efforts, talent, and time. People who have these skills generally are more successful than those who do not. The portfolio becomes the student's working directions, guide, timetable, record, and the hard evidence that work was done. A portfolio in our context is a self-built structure of guidelines and directions for getting things done in an orderly and logical fashion. Project structure and content is guided, designed, developed, and modified in collaboration with mentors and teachers. It provides a process of setting short- and long-term goals and objectives. This helps keep students on track so that they do not get lost. How we get project work done has to be taught, learned, and practiced. Although we see the final product, it is the planning efficiency, adjustments, and modifications that students have experienced during the process of their work that develop character, confidence, and good work habits that ultimately may be more important than the end result. This experience is just as important to us as the finished outcome.

All this puts students in control of their own investigations, research, and learning experience. The portfolio also provides mentors and teachers with student accountability for evaluation and assessment. Students are responsible for communicating, through their portfolio, mastery of the process and content of their work. If they did the work, they need to successfully communicate it. You can balance and weight the importance of each area of the portfolio with points. You can set intermediate checkpoints or grading requirements. Students with fewer organizational skills need more checkpoints, opportunities for feedback, reinforcement, and points of accountability. Together with the students, build them in. Grading can take place at carefully embedded points during appropriate times throughout the year.

Portfolio structure, major due dates, rigorous expectations, and assessment timetables are similar and standardized throughout the mentor program at C-TEC. The structure of the portfolio is separated into three large sections. The major areas are

- *Design and Development:* The focus here is on planning, methods, and protocols.
- *Implementation:* Implement the plan and make the needed adjustments.
- *Communication:* This is the demonstration of mastery over the process period. This section focuses on the production of work products.

Each section has its own timetable, requirements, and evaluation or assessment points that have been created collaboratively. The major structure works well for most disciplines and projects.

The requirements may differ because each teacher may have different needs, intermediate structures, or other requirements. Teachers assign point values to these areas and then weight intermediate tasks with points within the specific areas. Everything the student completes becomes evidence of mastery and understanding of the planned project outcomes and a relic in his or her portfolio.

6

Examples of Mentored Project Experiences

◢ These projects offered many opportunities to explore specific science disciplines within the context of the same problem. It is up to the teachers to determine and define the specific curricular pathways on which they want to focus. However, many of the projects overlap these arbitrary discipline boundaries because the projects are rich enough in other curricular content and process to provide challenges in these other areas.

Many projects can become authentic and real rather quickly. Consider the following passage taken from *The Bodega Bay Navigator*, March 10, 1994. Jim Sullivan critiques another article in the *Independent* on the local feral cat controversy:

> The meanest moment in the article is the gratuitous hatchet job on the Bodega Marine Lab's Drs. Peter Connors and Victor Chow. Not only did Jeff [original author] fail to comprehend that these two full-time professional biologists are in possession of a sophisticated understanding of the subject, but in his zeal to discredit them he also ridiculed the exemplary work done by the Piner High School science students who executed the field studies. On top of that, the *Independent*'s fearless journalist entirely missed another really interesting success story, Dr. Chow's men-

torship of what I'm told is a really great high school science
program at Piner.

Not all of C-TEC's projects become public controversies. This typifies
some of the experiences that teachers are hoping students will have.
They value the authentic audience.

The letter's impact can be evaluated in a number of ways. First, the
students and the mentor created this study. It is obvious that the
research had an impact on the writers of both articles. Second, it
validates the students' work as important to people other than just the
students' teacher. Third, the project and lessons became real-world
stuff open to public scrutiny. It was scientific research with a political
and emotional side. Although the research was not at a publishable
level, it was authentic and began to answer some questions about feral
cat behavior. And finally, it was a 2-year study that engaged 10
students in a meaningful way and a real-world, open-ended, local
problem with no real right answers. Science came alive for them.

Without examining, creating, and implementing new models for
educational opportunity, this experience would not have been possible in a traditional high school. The problem was a science problem,
but it quickly developed a social and political side. The research
became the students' to own and theirs to defend. In collaboration
with mentors and teachers, they designed, implemented, adjusted,
and communicated it. They built it and took ownership of it. The
project became rich with opportunities for all ability and motivation
levels. This is one example of a long-term project done within a local
setting, with local community professions, that brought meaning to
the students' classroom experiences. Here are examples of other project relationships.

The University of California—
Bodega Marine Laboratory and Reserve

The Bodega Marine Laboratory and Reserve is a field station and
a satellite campus of the University of California at Davis. Its primary
mission is to provide facilities for research and education. The areas
of research range in scale from molecular to cellular and include

research at the organismal and ecological level. The reserve portion of the site covers 361 acres of coastal prairie, dunes, rocky and sandy intertidal, mudflats, and salt and freshwater marshes that provide opportunities for a variety of field studies. Studies are a mix between basic and applied research. The facility serves researchers, undergraduates, and graduate students.

Students have been able to design and complete research with the help of various mentors at the lab for the past 4 years. In this case, a relationship was developed with a single person at the lab that was personally interested in education of this type. He has personally facilitated what it has become today. Others at the lab have contributed in many ways within their personal comfort level. Occasionally, university students will work with our students. Some act directly as mentors, some help out for short periods of time, and others are there for advice only. Each contact is unique, and students are carefully matched with mentors and projects. In some cases, the nature of the research is something in which the researchers are personally interested but do not have the time to do themselves.

Today, 11 students work at the lab. Their project research is as follows:

- Ranking order and social structure of a harbor seal haul-out site—are there patterns?
- Predator/prey relationships between the introduced Atlantic green crab and the local indigenous crabs—are they prey or predator?
- Studies contrasting and comparing botanical freshwater seep communities with the areas around them—how unique and different are they?
- Do tidepool communities reflect and support island biogeography theory?
- Marine fouling communities and paint toxicity—are there environmentally safer paint additives to prevent fouling?
- Two students are using the current molecular biology research being done at the lab as the subject of a video production to teach the techniques that scientists use to answer questions on the molecular scale.

Most of the expansion and growth of our program was handled informally as trust was developed and protocols were established. Parents and students are always asked to come for a tour of the facilities before research begins. Hazards and limits are identified, defined, and discussed. Every effort is made try to match the projects with the family's ability to support the student. If a student will be leaving campus and driving to the lab, appropriate district forms will need to be signed by the parents. C-TEC does have two cold water aquariums in the classroom that, in some cases, reduce the student's need to be at the lab, thereby reducing transportation problems.

The Broadmoor Project

This was a mentored project that included an entire honors biology class. The scope of the project covered a large content area, and it became the focus of class activities for 75% of the semester. It was considered an educational experiment. First, let me describe the project. Broadmoor North is a 14-acre parcel of land that our school district purchased as compensation for covering over wetlands on a building site for a new city high school. It also was purchased as a condition of the issuance of a building permit for the new high school for the district. The high school building site included wetlands of questionable habitat value. The Army Corps of Engineers issued a permit for their destruction only after the district agreed to purchase the Broadmoor site as compensation. The Broadmoor site contains wetlands and uplands. A portion of the Broadmoor wetlands are considered unique and include rare vernal pools. The Army Corps also required the district to draft and implement a 5-year monitoring and vernal pool enhancement plan. That is where C-TEC came in. Rather than district-hired consultants doing the required biological and hydrological studies, C-TEC, as an honors biology class, would do it.

The site is very pristine in comparison to the land around it. It contains two rare or endangered species, one plant and one amphibian, and there is a variety of native wildflowers and grasses on the site. It also contains a broad age and size range of valley oaks. Around this area, there are many large oaks but very few medium or small. Many of the younger generation oaks have been cleared for pasture or farm land. Broadmoor North is a small piece or relic of the original oak savanna that once covered the area.

The permit required a plan to improve the aquatic habitat value for the rare and endangered species. Baseline biological data had to be gathered as a control or comparison to measure changes in site biology after the proposed changes in hydrology are made. The idea is to increase the flow of water to the site, thus increasing the size and number of the vernal pools, which would increase habitat for the rare and endangered species in addition to other wetland organisms.

The bottom line is that the site is rich in biological and hydrological field study opportunities. The work had to be authentic and completed at a professional level required by the Army Corps.

The class indirectly and directly worked with a wetlands specialist, who oversaw the work. Other consultants included a hydrologist, a botanist, and an ecologist, who helped the class daily with experimental design, standardizing methods, data collection, and statistics. There were also two community volunteers—people with special interests and knowledge about the site—who added their personal expertise and a historical background. Research and baseline data were needed to meet Army Corps conditions. Other areas of research were identified as worthwhile even though they were not necessary for the report. Research tasks were identified, and students were given their choice based on their interest. Groups were then balanced, and students developed individual responsibilities within the larger research projects.

First, students, along with mentors and teachers, were asked to design and develop methods within the portfolio format to meet project requirements. These were reviewed by mentors and implemented and revised as necessary. There were many revisions because some projects worked well and others did not. One major project, hydrology, came to an end simply because we had a dry winter/ spring with nothing to monitor. Researchers in that group made a shift to soil testing, which turned out to be of real interest. Vernal pool soils are unique. Increasing the amount of water may not increase the size of the vernal pools if the water perks away. Soils are an important part of the structure. The soil data brought up questions about the potential expansion of wetlands. Mentors and teachers taught an embedded biology content as necessary to understanding the nature of research and the site.

Liability waivers were created for student drivers, and for a few weeks, the class traveled to and met at the site. C-TEC has a modified block schedule that allowed the blocks of time necessary. Students created brochures to help educate the neighbors, and they gave presentations to the Sierra Club and other organizations. A survey group completed a grid on the site with benchmarks and markers for relocating study sites, and an aerial photography company donated photos and other mapping services. The local paper featured the project in an article.

The project was very worthwhile, and the students will continue it. Now that the project is well-developed, it will not need as much class time to continue the project.

Nonpoint Pollution Project

Nonpoint pollution is roughly described as any pollution that is washed into the storm drains from surface water runoff and is not treated or removed at a sewage treatment plant. Some professionals or agencies estimate that as much as 50% of contaminants in city, urban, rural, and agricultural wastewater comes from nonpoint sources and is not treated or removed before entering natural waterways and other bodies of water. This includes runoff from streets, parking lots, and building surfaces.

This project was originally introduced by a parent who happened to be an engineer and owned an engineering firm in town. He knew the Environmental Protection Agency was soliciting grant proposals for various research and other activities related to nonpoint water pollution. The $90,000 grant was designed to provide opportunities for cities to develop a nonpoint pollution mitigation test site and a public education plan to reduce nonpoint pollution. Ultimately, the grant was cowritten by our city's Department of Public Works, the Regional Water Quality Control Board, and our school.

Mentors from these agencies supported and guided all aspects of the project. The mitigation test site, with its related technology, was installed and tested at Piner High School. Surface runoff from our site was trapped and treated in collectors installed on our site. The public education component was designed, developed, and implemented with the help of a local TV station and water agency. Student groups in a project class did the science and some of the engineering required by the project. Monitoring was implemented by gathering data needed to rate the effectiveness of the technology. Students started an education group to inform other students at the elementary schools. Others designed and produced brochures to be included in local water bills, and a nonpoint pollution video was produced and shown on local TV. All this work was structured within the portfolio format.

Summary

C-TEC's mentor program is just one kind of problem-based learning program. I am aware of other specialty programs that focus on the Westinghouse Competition and Science Fair. Students are engaged in original research, which is entered into this type of competition. Many of them find outside long-term mentors and laboratories because their home schools rarely offer the environment for research at the level of complexity required for the competition. In one school's case, it formed a specific class for students involved in this competition. Also in the problem-based realm, science fairs, in one version or another, have been around for a very long time. There are pros and cons to all these programs. The Westinghouse Competition may not be appropriate for all students within your classes. Science fairs have specific rules and timetables that may not fit within your curricular needs. Management of large numbers of projects can be a problem, so establishing a comfort zone within your teaching style is important. There are no singular right answers that are universally suitable. These are just examples of what has been and can be done, but there are certainly other creative ways to engage students in problem-based experiences.

7

Other Real Science Projects

◤ The following projects or problems work well in engaging students in the scientific investigative process. These investigative studies can be accomplished both inside the classroom and outside. Most do not require extensive scientific background because inquiry and analysis are the main principles. Your approach to class management needs to be well planned. The size of investigative groups, timeframe for data collection, assessment and evaluation outcomes, due dates, and backup plans for failed projects or absent students need to be considered carefully. Also, you will have to be involved personally with every study. You will become a working member of each group as the students run into problems and hit sticking points. Anticipating problems helps to answer student concerns and questions ahead of time. I have seen these projects used with a 24-hour time limit in a college ecology class. Setting concrete time limits focuses students, concentrates effort, and also provides a clean closure point.

Investigations are carried out with a minimal amount of materials and equipment, making them suitable to many classrooms. They offer opportunities to practice good expository communication skills, and they require statistical analysis. Content can play a big part, or it can be minimized to focus on investigative techniques.

Some teachers feel insecure with the open-ended nature of these problems. These projects allow students to investigate their own questions. With no right answers, teachers have to think on their feet, anticipate problems, and be vulnerable to not having an immediate answer. Confidence comes with practice. If insecurity is a pr͟, specific studies could be identified ahead of time and offered to thͅ students. This would provide teachers with greater control.

Each step in these investigations could be assessed and evaluated separately and include an individual timeframe. Each section of the investigation could be considered a single unit or module with its own timeline and outcome. The structure of the problem investigation and specific learning classroom modules might look like this:

I. Scientific Questioning

Activities center around what makes up good scientific questioning and how to turn it into a working hypothesis.

II. Experimental Design

Establishing methods to collect data and gather information to help refute or confirm the hypothesis is the focus of this module. Students collaborate to create a recipe or methods for data collection. A final test for the methods is that they need to be able to be followed or repeated by any other student—especially one who may not be familiar with the study. It must be replicable.

III. Conduct the Research and Make Adjustments

Students put their method to work, make adjustments as necessary, and collect the data or information to answer their questions.

IV. Data Analysis

Organizing and summarizing data is done in the classroom. If computer programs are available, students may need to be tutored in their use.

V. Communication and Scientific Writing

Traditionally, scientific expository writing is a specific skill much different from other writing styles. Saying more with fewer words is one goal. There are many models for a format. Most scientific journals offer examples of technical scientific writing. Saving past studies gives

ood and bad. It may be useful to require a
entation for critique before the final work is

, and outcomes vary with each section. Tight
required with one class and not with another. Each
te.   ed to add and adapt his or her own style to create a
perso.   fort zone.

Organisms Outside the Classroom

Field Studies

One school's field study site is located next to an acre of disturbed oak savanna, which contrasts an area next to it that is covered with nonnative trees, mostly eucalyptus. This plot also features a seasonal stream. Part of it is very natural, and part is a channeled creek. Most of the studies involve ecological concepts and relationships, but soil and water chemistry also have offered students study topics. Birds, insects, aquatic organisms, weeds and wildflowers, and grass and shrubs become the focus of fieldwork. To begin, all the data collection is limited, in most cases, to measuring and counting. This reduces the essence of the research to the basics of finding good, "do-able" questions; experimental design; and simple data collection and analysis. The steps are summarized as follows.

Questions

Classroom time is spent talking about the nature of science and discussing what makes good science questions and which questions are more philosophical in nature. Students then move on to talk about science versus faith and the steps of scientific inquiry. Students are given a written assignment asking them to reflect on their personal interpretation of science, the purpose of science, and how that may differ from faith.

Sometime during class, students are broken into groups and asked to tour the study site collecting 50 good science questions about the ecology or organism relationships at the location as they go. Once back in the classroom, out of the 50 questions, each group is asked to present three of what they think are their best, most do-able questions. Those three are placed on the board with the others. Sometimes, one of the questions is used to create a model of a working hypothesis, experimental design, and identification of investigative pathways.

Many students have problems getting 50 questions and come up with only "identification"-type questions that will not work for a study. At this point, you may want to discuss the types of questions that could be answered within the limits of equipment and time constraints in an effort to limit frustration. Other times, you may want to let them come up with poor questions and talk about why you cannot research them. The following questions are examples of "doable" questions.

1. Are bees more attracted to yellow flowers than blue flowers?

The question was turned into the hypothesis, "More bees are attracted to yellow flowers than blue flowers." Students set up study sites and experimental designs with equal numbers of both colored flowers. They considered the time of day and weather conditions when writing up their experimental design and data collection techniques. A few of the students gathered background research on bees and flowers, whereas others sat and kept counts over various times during the day. As side questions, some students in the group decided that they wanted to collect data on other insects and flower preferences as well. After 3 days of data collection, students came together to write up their results, put their data on a spreadsheet, and graph their data. They followed a scientific writing format and presented a short seminar on their work for peer review.

2. Do eucalyptus leaves inhibit weed growth or germination of seeds?

The students' experimental design consisted of gathering fallen leaves, grinding them up, and mixing various concentrations of the leaves with potting soils. Controls and duplicate studies were also added to increase the reliability of the results. Fast-growing seeds were planted, usually a variety of wildflowers, in the soils. After time had passed, an assay was done on the planted seeds. As with the first question, a paper was written, followed by a seminar and peer review. Because they did not use the same seeds from the plants actually at the site, validity issues were brought up by students.

3. Are there a greater variety of plants in the creek or on the land near the creek?

This was a tough one. The students had to decide how to pick study sites and eliminate variables. They also decided to ask other questions about greater or lesser numbers of plants and ratios of plants to one another as they moved away from the creek. Students focused on three related questions using the same transected and quadrated study areas. They were under the impression and formed the hypothesis that moving water would mean more plants and a greater variety of them. Because of the size of the project, frustrated students from less successful studies were brought in.

This group had problems picking a study site. Defining an "average" study site in order to limit other variables was giving them trouble. Part of their assessment and evaluation included identification of flaws in their studies after they finished and adjustments they would make if they repeated the study.

After working with a wide range of classes and a variety of interest and motivational levels, most teachers will find this project challenging for all students. The content may not always be complex, but it can be. Many students research the background material related to their studies, which were very open-ended with no right answers. The answers were only as good as their studies. From the teacher's standpoint, this was a no-mess activity. Rulers, tape measures, clipboards, stakes, nails, string, and rope were all that was needed in most cases. This specific class did have some simple water and soil chemistry test kits that were used occasionally. The true topic was science process.

Effective communication, either the written work or oral presentation of their projects, becomes important. Fellow students are quick to find weaknesses in the studies or presentations. Transparencies of graphs can be made and used in the presentations. The math was critiqued as to its appropriateness, quantification of data, and sample sizes.

Human Anatomy and Growth

Another way to have students experience the scientific process by simple counting and measuring is to follow the same procedure as with the field studies, but focus students' questions on growth, anatomy, and size within student populations. Average sizes and growth rates between classes make up most of the questions.

Some students get into foot, hand, and head sizes. More imaginative students may begin to investigate genetic traits such as eye color, attached or unattached ear lobes, or other traits usually covered in a biology textbook. Once you begin to build a catalog of past studies, students can refer to them, build on them, and modify or improve them. Growth rates between age and class groups become mathematical management problems. Sample size and random (there are scientific protocols that define what random is) versus haphazard selection of individuals within the study also present interesting problems in bias, assumptions, and overall study design.

By the time they finished, most students had a greater working understanding of how science is conducted. The concepts and principles of the scientific method are ingrained from experience, not just textbook study. Controls, variables, biases, and assumptions become important considerations during their studies rather than words used in a fill-in test. These studies set the stage for many later activities and further reflection.

Organisms Inside the Classroom

A good way to start this section is to give some examples. Many questions and investigations can be generated from student encounters with living things. These can become class, group, or individual investigations. Consider the following:

1. After a trip to the tidepools and a discussion on the sea urchin fishery, a student was looking at the aquaculture potential that urchins offered. The nutritional requirements seemed to be a big problem. Where are you going to get enough sea algae to feed them? A student wondered if urchins would eat other things, and if they did, would those things be cost-effective in an aquaculture model? We were lucky enough to have a refrigerated aquarium in class. After some urchin

background research, the student set the tank up and brought in the urchins. He selected various food types, and over four weeks of testing, his experimental design began to gather data.

2. As part of a "Salmon in the Classroom" project, students went on a couple of local creek walks learning stream ecology as part of the project. Various native species of creek life were netted and collected (they did have a collector's permit, but in many cases, a fishing license will do) and returned them to the classroom for observation. The group decided to set up a freshwater tank to answer questions on creek food webs and chains. Unlike many fish you are able to purchase, the creek fish had a ranking order and appeared very wild, keeping their natural behaviors.

3. Shore crabs in a marine tank became the subjects of a feeding study. This study was based on the crab's size in relation to its ability to feed with other larger or smaller crabs around. The student was also able to compare and contrast two different species in the same study. This study was done another year with sea stars.

Experimental design challenges students in ways that are truly unique. Eliminating bias and identifying assumptions in science investigations are high-level thinking skills. All of these studies became quantitative, and spreadsheets and graphing became important in summarizing and communicating their results.

Students were totally responsible for the care and feeding of their organisms, and the organisms were returned to the wild after the completion of the study. Seawater was brought from the ocean and minimally filtered. Ambient water temperatures were fine. Some water was exchanged over the duration of the project, but the short-term nature of the study and the organisms' return to the wild limited water quality problems.

In these examples, students always seemed to end up with aquatic organisms, but similar scenarios could provide research and investigative opportunities. Some students come up with their own questions and studies. Others require triggers and clues, and some were just given a study to do. Some can be used as class or group projects. The mathematical opportunities range from spreadsheets and graphing to complete statistical analysis with sophisticated computer programs. Most of these projects are based on ecological or organismal content. Any aquarium project could become a water quality or water chemistry problem. Water chemistry changes over time with organisms living in it. Many chemical test kits are available at local aquarium stores. Respiration rates, gas exchange, and temperature/behavior studies are other areas of potential investigation and research. These problems can be as rigorous and complex as the students can handle.

The Aquaculture Project

To the Teacher

This exercise was one that was originally given in a marine biology seminar class. It is really a paper project, but because we are asking students to produce an authentic plan, it falls in this category. This project may not be appropriate for some geographic locations. However, I know of Talapia (an African freshwater fish) projects in the hot, southern California inland areas. It could include both fresh (freshwater hatcheries) or marine water species. You may also want to consider aquarium exotics. When it was assigned to the college students, the directions were simple.

Pick any aquatic species that may have market potential and do a feasibility study for its aquaculture practicality under local conditions. Local native species may offer more resource possibilities. The college students had a semester to investigate and present an authentic plan. Students were to consider all the biological, developmental, and marketing issues and had to apply them to the local setting. There were also bureaucratic protocols, on all levels, to consider and all the logistical problems associated with any business. The student plans eventually became the property of Sea World San Diego, which became an authentic audience for assessment and evaluation.

The individual college students were asked to be responsible for all aspects of the project. In the secondary school setting, the project has many areas in which students could specialize.

I. The Science

    ◪ Developmental biology: Sources of brood stock and related development

- Reproduction and genetics: Maximizing genetic and reproductive potential
- Grow out: Biological issues once juveniles are released
- Disease: Preventive medicine and treatment
- Nutritional requirements: Cost/benefit analysis and food preference
- Water quality and environmental conditions: Water chemistry and habitat suitability
- Containment and predation problems: Containment design and maintenance and security of habitat

II. Bureaucratic Issues

- Government: Permit acquisition, zoning/land use, health and safety, sewage discharge, Fish and Game regulations, marketing regulations
- Marketing: Feasibility study, transportation
- Accounting: Financial planning and projection of a financial model

III. Project Management

- Organization and leadership: Producing the study and management of specialized areas

Ideally, this outline is produced by students themselves as an outcome of brainstorming the project. Depending on the ability levels and backgrounds of the students, this project can be completed totally by students or heavily structured by the teacher. There are any number of ways to structure this project. Outcomes and timetables could be created by the teacher to fit into the school calendar or constructed in collaboration with the students. Checkpoints and deadlines for assessment and evaluation of the project may be built in along the way. There are many structural examples within other projects that may work, or you can modify them to suit your needs. Finding or creating an authentic audience or recipient for the work may add to the importance of the eventual outcomes.

Resource acquisition and location can be a problem. Many governmental agencies have guidelines for businesses such as this. The state biologist is a good place to start. The local college or university may be involved in aquaculture projects. National Marine Fisheries or your state's Department of Fish and Game may offer help or other resources. Creating a list of potential sources or requesting materials ahead of time heads off student frustration. Once you have done this project the first time, it becomes easier the second time. Keep and bind student examples and create repositories of background resource materials.

The Seed Project

To the Teacher

This activity was developed and designed as a semester's research investigation. It was intended as an introduction to research, experimentation, and the scientific method. It is very much like the organismal studies discussed earlier. When I use this problem, I require more complex questions, research, and investigation than the organismal studies. It is also done at home or usually outside the classroom. It was used in the fall semester in California but may be more appropriate for the spring semester in other climates.

We did not produce student handouts on this project but developed them spontaneously on the chalkboard. In some settings, or with younger students, it may be in your best interest to design a handout to send home to parents explaining the project and its expectations.

We started by bringing in a variety of seeds and asked the students for 25 questions about the seeds or plant growth that they could answer using the seeds. Some of the questions could be answered within the constraints of the classroom or home. Standard responses included questions about trying different fertilizers, soils, or water. Other questions dealt with light and dark thresholds, gravity, or planting proximity. One project exposed plants and seeds to dry ice, CO_2-enriched environments trying to enhance growth. Interesting questions help motivate imagination and creativity, and they serve as the beginning to a science investigation.

Once workable questions have been selected, students develop preliminary methods that they think will help answer their questions. Once implemented, they usually need to be adjusted as the project continues. This includes standardizing experimental conditions, data collection, and measurements; and analyzing, interpreting, and synthesizing results. Analyzing the results includes mathematically analyzing and organizing data for communication of the results of the project. One of the bigger problems the students encountered was how to measure growth. Weight, size, volume, and surface area became interesting choice options. Scientific writing formats are much different from most of the writing styles to which they have been exposed. If scientific writing is required, you may need to teach it.

Assessment and evaluation can be structured as motivation to maintain the project over the entire time frame. In many cases, it is the longest assignment they have had in school. Surprisingly, few fail to complete the project. Most students did follow through and finish the project.

The first few days will be spent organizing team or individual goals and gathering information. Students will need to break down tasks into identifiable categories or subgroups. Some projects may require more than one student. The first tasks for students will be to brainstorm, gain background knowledge, and design and develop strategies and methods for the project investigation. An outline of these tasks will be used to construct a timetable for completion of team and individual work. Oral presentations provide peer review opportunities and help provide a forum for the exchange of ideas. This planning portion of the project can be an intermediate outcome.

There are many ways for students to be kept accountable for assessment and evaluation purposes. You can schedule "work in progress" presentations using a variety of demonstration strategies. This helps keep the students on task and focused. The final outcome becomes a scientific paper, oral report, and personal project journals and notebooks.

Separating Fact From Fiction:
The Product Test and Rating Project

To the Teacher

This is an introduction to the scientific processes activity. I have heard that this assignment has been around in one form or another for a number of years. This is my version. It makes a good first assignment to work on the students' perception of what science is. Because it lacks specific science discipline content and concepts, it allows students to focus on the pure investigative and research aspects of the scientific method. It is very open-ended and allows students to direct their energies toward a process of problem solving without needing a great deal of knowledge background. Once the methods are created, it can be an in-class or home assignment. Some teachers may feel uncomfortable spending class time on classroom activities not more directly related to content goals. We spent a good deal of time discussing experimental design and data analysis. It also makes a good classroom activity for shared peer review with team presentations at various points throughout the process.

Introduction

The true quality, or lack of quality, of many products or brands is shadowed or hidden in packaging and slick advertising. How do consumers evaluate product effectiveness or performance? Usually, decisions are based on perceptions, not facts. Good science is based on a procedure that answers science questions with reliable facts based on observation and experiments. There are many versions of similar brands designed and developed to do the same thing. How do consumers make their decisions on what to buy or which product is most effective? This project is designed to improve the student's ability to be a better consumer while conducting a scientific investigation that evaluates a chosen group of brands that are supposed to perform the same way.

Procedure

All products or brands claim to do something or perform in some way. In this investigation, you will choose a group of products with similar claims that interests you and contrast and compare their quality or the ability of the products to do what they claim to do. To do this, you will have to identify one or several qualities of these similar products or brands and design experiments to determine which product demonstrates the qualities more successfully. All performance qualities or factors need to be turned into some numerical value or scale of values for contrast and comparison. You need to rate performance factors such as speed or effectiveness. How you do it is the creative part. Develop your tests so that they are reliable and produce/provide the correct data necessary to answer your questions. Finally, you will need to consider price. If a product is high in quality and low in price, it is deemed a "best buy." Again, you can present these findings using math.

Use the following checkpoints to structure your investigation, and keep yourself working at a pace to finish.

Evaluation Checkpoints

This is an exercise in following directions, planning ahead, and meeting deadlines in addition to your scientific investigation.

1. **20%** Identify the products you propose to test. Write down your predictions, and define the factors you will examine. What are your questions about the product? Create rough drafts of a formal introduction and the experiments and methods that will help you answer your questions. Identify your numerical value system for evaluating differences.
2. **10%** Gather materials you need to do your investigation. Test your methods and adjust your experiments as necessary. Create your final introduction and methods (include rough data sheets).
3. **10%** Complete your investigation and gather your rough data. *Remember, the more data you gather, the more reliable your conclusions.*
4. **10%** Summarize your data using appropriate graphs and write a rough draft of the results.
5. **10%** Write a rough draft of a discussion. The discussion will include the following:

   - Reflections and critique of the results
   - Critique of your experiments and your methods: Did they answer the questions you wanted answered?
   - Price/value comparisons
   - Conclusion

6. **20%** Final interview: teamwork; planning ahead (use of class time); and degree of difficulty, challenge, and effort.

   There will be a bonus and extra consideration for those who identify and explain the science in the way their product performs. I will give you hints if you ask.

7. **20%** Evaluation of final paper.

Evaluation

Each checkpoint will have a point value based on the following:

- ◪ Did you follow the directions and complete the checkpoint task?
- ◪ Was your work presented in a neat and orderly way?
- ◪ Did you take suggestions and revisit and make revisions on past work?
- ◪ Maximum points for those who only need minor corrections and are ready to go on.
- ◪ NO POINTS WILL BE AVAILABLE FOR LATE WORK UNLESS *PRE-ARRANGED* FOR GOOD REASONS.

8

Defining the Outcomes
Problem-Based Assessment and Evaluation Philosophy

The three aims of assessment are demonstration of competence, feedback to students, and feedback to staff. Again, feedback and assessment at various times throughout the process helps avoid misunderstandings and "end-of-the-project surprises." Assessment can create important and useful information. Well-designed performance standards built into assessment instruments can enhance and validate the entire curricular experience. In many cases, the assessment will define the student's performance path because many students use the assessment and evaluation standards as a way of determining the amount of time and effort needed to receive the "grade" reward. Some still will define learning something in class as the reward unto itself, and many fall somewhere between the two. In any case, moving into a problem-based curriculum means a rethinking and relearning of instructional assessment and evaluation strategies. This may be necessary for teachers, students, and parents. Any confusion can cause students and parents anxiety. Designing, developing, and communicating well-designed and organized assessment and evaluation tools and instruments can enhance effectiveness and limit misunderstanding and problems.

In problem-based learning, assessment and evaluation must be based on the application of appropriate knowledge in correct context to practical, relevant situations, as well as on the student's ability to demonstrate the ability to respond, manage, and solve or resolve problem-based learning activities. These criteria need to include and consider both individual and group expectations.

Individual assessment within groups can be done successfully within a portfolio format. There may be some duplication of materials, but the individual should be able to document his or her unique contribution to the group product. Assessment objectives and goals are based on collaboratively defined beginning and intermediate outcomes. We are concerned not just with the end of the project, but with intermediate outcomes, the smaller steps within the process. Possession and retention of knowledge should be balanced with overall demonstration of problem understanding and management agreed upon at the beginning of the project, and also whether the information is correctly applied, in context, to a problem. This may include self-assessment instruments developed within the project.

An example of this is a required weekly learning log completed as part of the self-assessment. The learning log idea asks students to communicate reflections of what they have done and learned. It also asks them to project their plans for future work and to ask for whatever help they might need. This is a good way to get important feedback for troubleshooting potential problems.

Evaluation requires new strategies. Some outcomes are hard to quantify. A pass/fail system is less likely to promote competition between students or disrupt group dynamics or teamwork, but in many schools, this simply will not work. Parents and potential colleges want students to have grades. Some programs give an "A", a "B", or no grade (incomplete), and those that do not receive a grade are given a specific time frame to bring their work up to "A" or "B" standards. Others complete a rubric with very concrete criteria that need to be completed in order to have their project considered for assessment and evaluation. Did the student provide a bibliography and resource list, and complete a report in the proper format? These are points that may not reflect on the content or process of the work.

All of these issues are built around the personal preferences of the teachers and their professional reflections on what would meet their students', parents', and administrative needs. Going back to literacy, the levels of the term *literacy* vary. What level of literacy is your goal, and how are you trying to assess mastery of it? There are many good reasons to create, in advance, problem-based learning activity models and outlined plans of the learning experiences, including timetable preparation, implementation, development of knowledge and understanding, staff and student plan of assessment and assessment instruments, and the final demonstration of understanding. These plans, together with the expected level of competence and standards for individual students or student groups, establish a teacher's credibility and control over the learning environment. They should include both group and individual expectations, goals, and objectives. The rubric— an itemized checklist of behaviors and activity outcomes—that a student is expected to exhibit will help eliminate communication problems between teachers, students, and parents. Projects and problems that are repeated can be modified from previous experiences. As mentioned earlier, "The assessment tail wags the curriculum dog." They must be considered equally. Although there are a number of concrete examples of assessment, evaluation, and grading elsewhere in the text, it is useful to explore a range of ideas and thoughts on the subject.

The goal of assessment in most systems is to have students produce information for the purpose of assessing competence, performance, and achievement while giving feedback to students and providing feedback to staff on the effectiveness of the specific program. Assessments are usually criteria referenced and relate to educational objectives and goals. Inferences are to be drawn on assessment results and may reflect success or failure in meeting certain expectations. They also must be considered a learning device in and of themselves because they are an intellectual puzzle that students are asked to solve regardless of the testing strategies.

The awareness of what student performance and behavior represents is not the sole province of the teacher because students themselves may develop similar or very different reflections of the results of any assessment technique. This notion may lead to self or peer assessment and evaluation. The timing of these techniques in any activity may contribute to or reduce the effectiveness of a student's experience. With planning, these techniques can enhance performance and help students practice and adopt assessment skills that serve them well in the future as well as the present.

Classifying strategies can be messy, and there is overlap and integration between models. However, a starting point is to break these strategies into three distinct areas: content, process, and outcome.

Content: Information and Knowledge

Content assessment and evaluation are concerned with information, knowledge, concepts, and principles that students have acquired in their memory banks and can bring forth by recognition, recall, or association. Many times, answers are present or cued, the correct answer may be in front of them, and students will have a choice to make. This is not a pure definition, and some content evaluations may ask the student to process data by identifying similarities or differences, synthesizing and integrating data, and analyzing or abstracting information. It is usually subject based and teacher centered. Vocabulary recall, in a variety of forms, is very often the most important element in preparation for student success in this category. Concepts and principles may be recalled from memory rather than personal experiences with their relevant application. Fill-in, multiple choice, and true-false formats are typical of this model, and some ask for some form of descriptive essay response.

Process: Methods and Techniques

This type of assessment tool is concerned with the learner's ability to structure a problem-solving framework, use information to solve problems, and evaluate information or data. They may be asked to demonstrate a process or protocol that could be applied to a specific problem. Content retention is not measured directly. Process-testing frameworks may be authentic, simulated, or role-play. They are designed to allow students to demonstrate reasoning, inquiry, problem formulation or analysis, and interpersonal and decision-making skills within a single subject area, or create integrated and interdisciplinary responses. The format usually includes applying these skills acquired in previous activities to new situations requiring the same or similar skills. Formats may or may not include cues (hints or no hints) on the level of the students. It is possible to assess the content by the way it is applied to the process. This method is generally associated with problem-based learning. Expectations can include a content-based component; however, there may be limits to the effectiveness of this model to measure content.

Outcome: Intellectual and Concrete Products

Outcome-based assessment may not be considered a test in the classic sense. The outcome of any given activity can serve as an evaluation tool for problem-, process-, and content-based educational objectives and goals. Typically, the student has produced a product, and criteria have been established to gauge the effectiveness or success of the student response. They apply knowledge to practical situations, manage and solve problems, and produce evidence. This style may include intermediate process or content outcomes. Students are producing original responses to educational triggers that are very different from consuming and recalling information. Group or individual outcomes can be structured. This is a bit of a role reversal. Instead of consuming information, students are producing a product that may be new and original knowledge or a new application of knowledge.

A range of problem and solution fluency within a flexible framework of approaches is what problem-based learning is all about. Originality, creativity, and the ability to contribute to and use group support within any educational activity is unique. These qualities must contribute to an overall reflection of performance. Success in this model is built over time from a wealth of specific prior experiences. The inventory of prior experience and patterns of problem-solving recognition can be used when confronted with new or unfamiliar problems or situations. The pace at which these skills are acquired varies with the ability, engagement, interest, and motivation of the learner, and assessment needs to reflect this. Uncued and open-ended formats may work but may also have consequences that must be considered. Vagueness in uncued formats can cloud what the "questions/tasks" really are. Cued triggers may alleviate frustration or "test anxiety" as learners acclimate to a new learning style.

Relics of all assessment processes may vary from bubble-format answer sheets, reports and essays, artwork, speeches, or oral discussions to computer programs or video. There is a wide range of potential here. Within the problem-based assessment lies a problem. Many of the evaluation tools used may not be available to the student after leaving the program. Consider this when designing evaluation instruments to facilitate students' development of their own approaches to self-evaluation that can be continued outside the program.

Logistical problems may limit the optimum choice of assessment tools. After working through process problems or creating product outcomes for most of a semester, giving a content-heavy test at the end of the semester because of time constraints may not be appropriate. The assessment tool must produce information to measure the ability of the student to respond to the expectations of the curriculum. Creating valid and reliable assessment tools and evaluation techniques is done in relevant context and needs to be consistent with classroom practice.

Finally, correlation studies of relationships between assessment styles and their results are counterproductive. A true-false, multiple-choice test rarely would be a valid measure of many problem-based activities. Performance or application-based test scores; those associated with problem-based curricula; and other, more traditional assessment methods target different expectations, objectives, and goals and therefore produce different information. Specific validation work should emphasize the study or critique of problems or threats to the validity of score interpretation as it relates to specific curricular expectations, not the relationship with other curriculum styles or other assessment measures.

Assessment and evaluation are a high-stakes activity for students and teachers. Regardless of the method used, assessment will have an impact on teaching and learning. The nature of this impact, especially within any new curricular expectations, is not predictable. The politics of teaching require a high degree of thought, awareness, and preparation in the formation of assessment techniques. Students deserve fair assessment opportunities.

Neither more traditional assessments, performance-based assessment methods, or product outcomes provide the total answer. All clearly assess and measure different skills. Performance-based tests, done well, can assess skills that cannot be measured with traditional written tests. In areas where assessment of breadth of knowledge and skills are a large concern, the problem design time, testing time, and resource requirements needed for performance-based testing to achieve adequate coverage make this method very impractical. A well designed blend of methods may be a solution.

The Rubric Philosophy

A typical rubric defines the teacher's specific expectations for evidence of task completion and the mastery of targeted instructional goals and objectives; expectations for different levels of student performance; and overall demonstration of problem management and mastery. For the student, the teacher, and the parents, it serves as a guide for student performance expectations, and it provides guidelines for standards, assessment, and evaluation of student work within any curricular activity. For some, it becomes an itemized list of goals and objectives. The "classic" educational objective stated an outcome with some mastery demonstration or evidence. An example of this might go as follows: "Students will understand the concept of taxonomy by successfully forming a classification scheme for 25 hypothetical organisms." If they can do it, they have mastered it. Ultimately, it should define what a student should know and be able to do by creating a clear picture of the evidence a teacher will be considering as a demonstration of mastery of the content, content processes, and related behaviors within any instructional activity.

The successful rubric is able to quantify, define, and match all outcome expectations with concrete pieces of evidence for demonstration of mastery of goals and objectives. An expectation such as "works well in a team situation" creates problems because most of us know this when we see it but cannot quantify it. The content-based test works well because answers are typically very concrete, and questions are itemized and easily quantified. Mastery is demonstrated by the number of correct responses to the questions. Once leaving school, most of us rarely need to demonstrate job mastery by test taking.

Our mastery of our job or work is performance demonstrated in other ways. Most professionals have their job performance and competency assessed by methods not used in a school setting. They produce work products, reports, records, evaluations, and so on that are usually required to meet or exceed some quality standards and guidelines. Assessment and evaluation rubrics can begin to define these quality standards and guidelines in addition to defining evidence of the mastery of tasks within the problem-solving process.

In a problem-based/student-centered approach, the rubric can be a product of collaboration between the teacher, student, and parents. If expectations are driven by the school, school district, or state guidelines, the teacher could use them to connect to and construct the rubric. Either way, emphasis should be placed on clear communication of the rubric to students and parents.

In a typical problem-based activity rubric, you could divide the rubric into three assessment component areas. Assessment and evaluation would be based on concrete examples of evidence of mastering expectations within each problem-based philosophical instructional component.

Problem-Solving Organization,
Background Research, and Data Collection

Organizational skills can be demonstrated by creation of problem attack plans, hypothetical experimental designs, outlines of planning strategies, and ongoing progress updates. The learning log idea could work here.

A bibliography of all references and resources acquired may act as evidence of an information search. There could be a maximum and minimum number of any type of reference materials. Limiting or restricting students to the most current references or sources of materials keeps them looking at the most relevant knowledge. You may ask for file card summaries of information within the reference or the material they actually used.

Evidence of data collection applies to experiments or investigations where students generate new data, which all can be quantified. The amount of data collected often has a time component. The more time the students work, the more data they collect. Some data are harder to generate than other data. Teachers will have to use discretion here. This can be communicated through data sheets, graphs, oral presentations, or written reports.

Application of Information to a
New Model or a New Problem

Once the organization is created, the research is completed, and the data are collected, the data usually need to be interpreted, analyzed, applied, or implemented within a new activity paradigm. Something new is expected to be produced with the collected information, knowledge, and techniques.

This might be demonstrated in the production of an intermediate or preliminary model (written paper, rough draft, progress report or learning logs, or product prototype). Usually, this part of the rubric focuses on assessment of the process of problem solving, teamwork outcomes, and intermediate outcomes and adjustments. Adjustments and modifications to the problematic attack structure redefine the focus, direction, and path of the project or activity.

Individual students add their specific contributions to their portfolios and also keep a record of any group or teamwork products.

Formation and Communication of the Final Product

The rubric now includes the interdisciplinary expectations that are associated with creating the final product and the communication of the new information to an audience. Effective presentation techniques become the goals and objectives. A rubric must include and stress that any presentation is a "test" of students' mastery of the targeted subject content, techniques, and associated processes. The rubric must contain standards for these in addition to standards for the communication techniques for the demonstration of the discipline mastery.

The Practical Rubric

Let's create, design, and develop a rubric for a typical problem-based activity. We will consider the *Island Project* as an example. This activity typically covers a 3- to 5-week time span. It can be used in any science class and can be designed to stress or focus on biology, earth science, or even chemistry. It would be more difficult to match up to a physics curriculum. It is rich in content, and a content map could be constructed to match the activity components with textbook chapter content.

The following scenario is a hypothetical example of the development of an island project rubric. The rubric itself was produced during a class period in collaboration with the students. The outcome was typed and given to students as a timeline guide. A teacher has the option of asking students to share it with their parents.

The assignment was duplicated and handed out to the class. Students were asked to read the activity and be prepared to participate in the formation of a rubric. This school evaluated students every 4 weeks for a progress report, gave quarter grades every 8 weeks, and gave a 16-week semester grade. Ultimately, students' assessment and evaluation of the island project would make up 100% of a 4-week progress report grade.

Early in the period, students were told that they had the option of working in groups of no more than three students. However, they could work in pairs or individually if they or their parents desired. This served to eliminate arguments about group or team grades. They agreed to accept the consequences of team and group activities, good or bad. The responsibility of working in teams and the typical problems associated with team activities were discussed. Most notably, it was made clear that being less motivated or absent would not excuse individuals from completion of any project expectation or deadline.

Parents often expressed concerns when their student worked in a group and received a group grade. This strategy was a choice that each student made. A teacher might decide to assign individuals to groups, randomly assign students to groups, or use some other method to create teams. This would be included in the rubric introductions. Once the student grouping issues were decided, the groups signed up, and switching teams was not an option once the activity had started.

The groups were given the option to develop their own work strategies and divide the workload any way they wished. At this time, it was the teacher's decision to use the portfolio format as a management, assessment, and evaluation tool. All assessment of student mastery was based on the evidence in an individual's portfolio. The rubric started out by explaining the content, process, and technique goals and objectives. Next, it defined team or group ground rules and then moved to the concrete objects that made up the specific outcomes for assessment and evaluation. Many times, these are considered intangibles but can be important to parents and students. They also project a professional competence and control of instructional strategies and classroom management. Next, the portfolio contents and timetable needed to be developed.

The school calendar was written on the chalkboard. Students were told that the portfolio structure would help them manage their workload and enable them to finish on time. The teams agreed that it was a good idea to break up and define responsibility for specific tasks. Students could decide what topics they wanted to explore, and they also would be free to brainstorm any unique approaches they may want to take. For example, one student wanted to create a contour map of the island using a computer program. Another wanted to access weather and climatic data from a site near the island location on the Internet to better match the colonizing organisms to their preferred climates. She intended to use it to create a habitat scenario that reflected the authentic conditions at the site. This information was to be graphed and plotted within a computer program, and it would be very valuable to her teammates, who were researching the potential for biological colonization of the island.

A written management plan was a suggested addition to the individual portfolio as evidence of organization. Most students immediately asked for a few days of class library time. It was agreed that individual bibliography cards would be kept. Duplicated copies, written notes, and other references could be added to the separate portfolios in addition to summaries of reference information on notecards. The individual specific references would be added to the final group report. All of this material would become evidence of a rigorous information search.

As discussions continued, the teacher expressed the importance of the value of each step in completion of this assignment. The first three points of individual portfolio assessment were included to reward students who were able to pace their project workrate. If content retention of specific concepts and principles, facts, vocabulary, and so on are important to you, each could become a point of mandatory inclusion within the final product. For some, a more traditional test could be included for these curricular goals and written into the rubric with a point value at the completion of the project.

Island Project Rubric

Project Structure

Basic Ground Rules

- Students will have the option of working in groups of two to three, or individually.
- The instructional content and process goals and objectives are contained within the written assignment.
- There will be a 4-week timeline for this project. The grade for this project will constitute the 4-week progress report grade.
- Assessment and evaluation will be managed through the construction of a working portfolio and timeline. Its content expectations and specific due dates within the project timeline are mutually created by the students and the teacher.
- Students are expected to be prepared to work all period during class work days. This will be an individual assessment consideration.
- The individual portfolio will be assessed three times: at the end of the seventh project working day, the Friday before the last week of the project, and at the completion of project communication during the fourth week.
- Evaluation is based on a 200-point system. The first 150 points are available for the individual's performance and contribution to the activity (portfolio contents) and the final 50 points will be available for the group's performance. The final product, report, and presentation will be assessed for content and effective communication of project results.
- The final written report guidelines include double-spaced text, Times Roman font, and 12-point type. A scientific writing format and basic expository writing guidelines will be handed out in class. Students will be expected to use this style when writing their final papers.

Individual Portfolio Content Expectations (100 pts.)

First Check (50 pts.)

- Timeline and project calendar

- Group working plan and individual responsibilities (separate from the group plan)
- Individual bibliography (minimum of eight references) and summary notecards, duplicate copies of references, written notes, and other evidence of research activity
- Team bibliography

Second Check (50 pts.)

- Rough draft of individual contributions
- Written progress report

Third Check (50 pts.)

- Update of any additional individual contributions
- Self-evaluation and grade expectation. The self-evaluation is a personal statement that could include a reflection of personal effort, unique contributions to the group, or an explanation of project results.
- Copy of final report

Final Paper Assessment Considerations (50 pts.)

- Addressed all or exceeds content and process objectives
- Followed writing style and expository writing format
- Made an oral presentation that was organized, professional, and thorough

Grading

For this project, a mastery assessment and evaluation approach is applied, as opposed to other grading methods. If all the required items are present in the quantity requested, the students likely will be given a grade in the "A" or "B" range and related point values. Anything less probably will be given a "C" grade or below and related point values. The final grade will be based on the accumulation of points from the four rubric categories.

An analysis of specific grade categories might look like the following examples. These descriptions are a bit complex and could be simplified for a specific class or audience. "A" and "B" work is covered here, and you could extend this criteria to "C" work and below.

A grade of "A." All portfolio items are present. The individual portfolio reflects thorough research of reference materials, associated with the student's responsibilities and group plan and exhausting all of the school's resources associated with this project. This includes the class textbook and classroom and library references. In addition, references reflect off-site research acquisition and search efforts. Research work is completely analyzed, clearly documented, and ready for implementation with only minor changes to methods, procedures, or timetable. The skills and equipment required by the project can be mastered and acquired by the student.

Rough drafts of contributions and progress reports offer appropriate responses to project goals and objectives, and insight reflecting control over all aspects of the project. All project science content goals and objectives have been addressed. Coverage of all facts, concepts, processes, and general science content associated with this project are mastered, applied correctly, and demonstrated in the rough draft and final products. All of the knowledge needed to solve the project requirements has been considered and addressed. The self-evaluation illustrates realistic assessment of the positive and negative aspects of the student's performance. The student made good use of class time and is a clear asset to the group. Both the paper and the presentation reflect appropriate mechanics and format, and the written work contains few grammatical errors. The student has engaged in and taken ownership and responsibility for his or her project.

A grade of "B." All portfolio items are present. The individual portfolio reflects thorough research of reference materials, associated with the student's responsibilities and group plan, exhausting all the school's resources associated with this project. This includes the class textbook and classroom and library references. Research work is analyzed, clearly documented, and ready for implementation with only a few changes or suggestions to methods, procedures, or timetable. The skills and equipment required by the project can be mastered and acquired by the student.

Rough drafts of contributions and progress reports offer appropriate responses to project goals and objectives, and insight reflecting control over all aspects of the project. Most project science content goals and objectives have been addressed. Coverage of most facts, concepts, processes, and general science content associated with this project are mastered, applied correctly, and demonstrated in the rough draft and final products. Most of the knowledge needed to solve the project requirements has been considered and addressed. The self-evaluation illustrates realistic assessment of the positive and negative aspects of the student's performance. The student made good use of class time and is a clear asset to the group. Both the paper and the presentation reflect appropriate mechanics and format, but the written work contains some grammatical errors. The student has engaged in and taken ownership and responsibility for his or her project.

Problems With Rubrics

What can happen with a rigid project management scheme? Christmas and proms break up project inertia and momentum. No matter how you plan, illness, family problems, computer problems, holidays, missing resources, and so on can create the need to consider adjustments and plan alternative pathways because of these problems. It is always best to have answers and responses ahead of time for students and parents. No matter how well you plan for group or team instructional activities, there will be problems to deal with. Heterogeneous classes offer challenging management problems. Assessment and evaluation is not always based on rigid standards and may need to be adjusted, informally or formally, for the teacher's personal reflections of individual student potential. This is a very personal task and may be impossible to quantify. Ultimately, you want the benefits that the heterogeneous class offers—maximizing group learning strategies within various student motivational and ability levels—and an unlimited, challenging curriculum, teaching strategies, and methods for all learners. The problem-based, student-centered approach would seem to eliminate many of the complaints from both sides of the homogeneous and heterogeneous argument. However, assessment and evaluation do present some interesting problems for a heterogeneous classroom. Regardless, a collaboratively designed and well-communicated rubric serves every stakeholder's needs. It bestows credibility and adds professionalism to curricular development.

9

Curricular Accountability

Problem-based, thematic learning, like any other teaching and learning methodology or instructional strategy, comes under scrutiny from a variety of perspectives. All groups or individuals, within educational communities, from students and teachers to parents and administration, have their own scale for assessing and evaluating the validity of teaching and learning. This is done both informally and formally. Curriculum designers need to be able to justify and provide evidence of the suitability of curricular activities. If teachers are not creating their own curriculum, they are bringing it in from other sources. Whether it is produced on site or brought in, it needs to be evaluated for validity. Curricular accountability is at the heart of any validity and effectiveness scale where teaching and learning take place.

A specific example is used to illustrate and provide an example of how a problem might be examined for embedded content and process. In addition, I have looked at some of the general features of the problem-based pedagogy unrelated to content and process. These features are important management considerations when assessing the appropriateness of any curricular paradigm.

Content and Process Features

As described in the introduction, there are many standards, frameworks, college and university expectations, guidelines and philosophies, and testing and assessment instruments that address what should be taught in science courses and that measure what students retained. Some focus on specific vocabulary and content, whereas others focus on concepts and principles, and some include application of content. Few define or assess and evaluate pedagogical effectiveness. Instruments to assess pedagogy can be designed and implemented by teachers. Terms and phrases such as "less is more," "depth not breadth," and "hands on or minds on" further add to curricular confusion. I have seen some very poor hands-on activities that were, at best, very loosely related to any scientific principle.

Accountability to content, vocabulary, techniques, and processes may be important in some settings. Although teachers rarely are able to cover all the content in textbooks, most have specific content areas they deem very important. The table of contents in textbooks makes it easy to assess coverage. Problem-based learning modules make it a little more difficult. Accountability to the coverage of important curricular content objectives can help justify a change in teaching style, pedagogy, or instructional methodology. A few years ago, a book company, as part of its sales packet, included what it called a "CAP map." This was in California, and CAP meant California Assessment Program. The book publishers wanted to connect their textbook to the performance expectations and content of state testing. They wanted to show curricular accountability to the test.

Each one of the investigative examples in this book has a large number of possible pathways, each with its own specific subject or concept emphasis. As you read through them, most paths, themes, and topics are obvious. However, there always seem to be a few students who see a pathway you have not thought of and who want to explore it. The beauty of these problems is that modifications and adjustments can be made easily. There are many different spins that can be placed on the same problem depending on the content and process coverage that the teacher desires.

A curriculum writer can connect problem-based activities to concept and process expectations and coverage. Consider the following laboratory experience. It was introduced orally to the class and structured collaboratively on the chalkboard.

The Adaptive Characteristics of Intertidal Sea Algae

The Problem

A student made a statement during a trip to the tidepools along a rocky intertidal zone. She said that sea algae nearer the shore would be able to retain water in their tissues longer than those covered with water more of the time. Others questioned her statement. This trait would help them survive drying out when the tide was out. It is easy to turn that statement into a question. Are seaweeds in the high intertidal better able to resist drying out than those in the lower intertidal?

There is a range of many different species of sea algae within the intertidal zone. This zone is known as the *triple point,* where water, sea, and air meet in a narrow zone bordering land masses. Unique organisms have adapted methods of dealing with the stresses of life in and out of the water. They are required to cope with a large number of environmental conditions: wet and dry, hot and cold, fresh and saltwater, high saline and low saline, heavy surf, predation, and competition for space. Some organisms spend much of their lives in water and others out of the water. Desiccation, or "drying out," can be a problem, especially for sea algae or seaweed that have no mobility. One would expect sea algae in the higher intertidal zone, the area that spends the greatest amount of time exposed to air, to have adapted characteristics to absorb water quickly and retain water while exposed to air and the hot sun. Conversely, those plants lower in the intertidal zone, covered with water more of the time, would be less likely to exhibit the same characteristics. If this hypothesis is true, those plants in the middle intertidal zone would be expected to have absorption and retention rates somewhere in between.

The class problem is to design experimental methodologies to investigate this hypothesis and answer the student's question. Students placed each algae type in seawater overnight to fully absorb water. The next day, the algae were taken out of the water, placed in containers, and set on scales. Samples were measured for weight every hour as the algae desiccated. Water loss data, over time, were plotted and graphed. They reversed the process to measure the plant's ability to absorb water. Standard methods were developed to weigh and deal with the water clinging to the algae. Once data were collected, the plants were placed in an oven to completely desiccate them; once dried, they were weighed once again.

For the purposes of accountability, the major concepts and principles here are the scientific method, natural selection, adaptation, and evolution. The subjects are plant physiology, anatomy, and ecology. Specific content processes are osmosis, facilitated transport, transpiration, and behavior and function of cell membranes. In an integrated approach, the properties of water, solvents, solutions, ions, polarity, atomic bonding, and so on could be explored.

Without a source of sea algae, this project could be simulated. Elodea, an aquatic plant available at aquarium stores, could be used as a substitute. Picture this scenario.

Pollution Problem

Samples of elodea, which had been introduced to a large lake years ago, were brought to the school for analysis. Many times, unwanted aquarium fish and plants are dumped in ponds and lakes. Exotic organisms sometimes find favorable conditions and thrive. This was the case with the elodea. It colonized the shallow muddy areas around the lake.

In this case, the elodea in certain parts of the lake (54 square miles, 60 feet deep) appear to be dying. Preliminary microscopic analysis reveals plants with cell membranes that are pulling away from the cell walls, as well as cells that are shrinking in size. Samples of water and elodea have been taken from eight locations around the lake. Land around the lake is used in a variety of ways: Agriculture; a small, industry-based city; and a forested area border the lake. Three streams feed the lake, one year-round and two seasonal.

The cellular water loss, which is due to the mineral salts and osmotic balance, becomes a working hypothesis. It is the student's job to design methods to duplicate the conditions in the school laboratory and test the hypothesis.

The teacher's job is to create the mystery and the evidence and provide the clues or triggers to coach the students through the science and help manage each student's role within the project. Initially, a specific number of samples can be kept in water containing various concentrations of sodium chloride (table salt). Elodea cells will shrink at various rates in response to the different saline concentrations. Samples can be deemed as originating from a variety of lake locations, and students can conduct investigations of these samples. The high saline samples can be associated with either agriculture or the city, both potential sources of mineral salts. There may be other imaginative sources of pollution built in. The complexity of the scenario can be tailored to the student group. Students like mystery and stories.

The themes and topics of various educational pathways contain examples of various scientific concepts and principles. Scientific methods and experimental design are major components. Specific content covers the same areas as the algae problem with the addition of introduced species and nonpoint pollution. The same chemistry content applies here, and there are land and resource issues here also. Human activities within this simulation may include ethical and societal issues.

Summary of the Content and Process Features and the Characteristics of Problem-Based and Thematic-Based Activities

1. Important scientific concepts, content, and processes can be covered in depth and in the context of their application. These are linked together in an integrated and interdisciplinary style.

2. Most problem-based science activities, including the examples, can be linked to a math curriculum in a quantitative applied authentic context.

3. Theories and hypotheses can be examined and tested, and moral and ethical issues can be built in for consideration and examination.

4. Learning opportunities are open-ended, not just recipes to right answers, and students are free to explore and experiment within their own interest or learning styles.

5. Laboratory and fieldwork can be integrated within the curriculum.

6. Problems can be designed and developed to use the range of technology and laboratory equipment available at the site. Many "canned" activities are limited by lack of equipment.

Summary of the General Curricular Features and Characteristics of Problem-Based and Thematic-Based Activities

1. Teachers are free to build cooperative learning opportunities and offer leadership roles within problems. Multiple instructional strategies can be included within the same problem, and mastery can be defined in a number of ways.
2. Special student populations benefit because teachers can customize their opportunities, expectations, and roles within each problem. There is no one way to learn here.
3. Flexibility is a key attribute of this teaching and learning style. Most problems can be modified and adapted while in process.
4. An appropriate level or balance between teacher direction and more self-direction can be built in to problems as the student's experience dictates.
5. Various technologies, which range from computer programs such as word processors or spreadsheet programs to traditional laboratory equipment, can be included to enhance and complement the professional atmosphere that teachers want to foster. A portfolio-based management scheme, discussed in earlier chapters, helps assess and provide evidence of a student's effort and accomplishment.

10

Science Education Then and Now
Students, Parents, and Staff in the New Paradigm

◢ When students do not bring home their science textbooks or talk about their latest animal dissections, parents start to make comparisons between the science they experienced and the science their children are now experiencing. From our adventures in reform, change frightens and threatens students and parents. The student and parent perception of what science education is and the experiences students should have become an obstacle in the reform effort. This is especially true in classes full of college-bound students. If students do not meet with instant success, parents lose confidence, and criticism comes quickly. The parents' view of school is based on their experience, and if their student's experience is different from what they remember, there may be some concern or mistrust. A problem-based science program is not immune to this. Parents need to be educated in new teaching and learning methods also. As with any new model, both small and large, it has to be tinkered with, adjusted, and fine tuned, and we need the support of the administration to do this. Communication with parents demystifies and reduces potential problems. The best time for these meetings and discussions is before school begins or soon after. Talking with parents after grades or report cards come out is not a suggested strategy.

If students become just the *beneficiary* of a change in teaching and learning styles and not *participants* in the change, problems may develop. Students should have the opportunity to anticipate and experiment within the new roles you design and develop with them before they are expected to demonstrate mastery of any given problem. Whether you plan for it or not, they are participants in the reform process because you expect changes in their roles as students. If students have routinely viewed learning as memorizing information or taking end-of-chapter tests, discrepancies will exist between reform expectations and students' perception of their role. If a passive role has served them well in the past, you may be sending them confusing mixed messages because you expect them to be active learners now.

These were problems that seemed to come up with certain groups of high-achieving students. These were groups that saw school as a stepping stone to college. Intellectual work beyond memorization threatened their grade point average. The open-ended opportunities were perceived as time-drainers in their academic balancing act. Others were insecure moving into a new learning and teaching style. Jumping from a passive to an active mode while moving between classes or jumping between teaching styles within the same class may cause confusion. The appearance of acceptance of active learning may be just superficial engagement. Students need help in accepting expectations as being valuable and appropriate to who they are and where they want to go. For most students, this was not a big problem; however, when it did come up, and after talking with those students, we found it best to acknowledge their concerns and let them move to a more familiar educational approach. Planning ahead for a range of student acceptance can help teachers avoid sticky situations when classroom changes are implemented. It is therefore reasonable to include students—those expected to change—in planning, implementing, and evaluating the new curriculum.

Parents can be participants as well. Remember, they often rate their children's and your teaching success based on student grades. Sometimes, grades measure either your ability to teach or your program rather than the student's true performance. Within some households, grades sometimes become more important than what students learn at school. Grade pressure from home is not supportive of reform. Parents who are active, interested, and engaged in facilitating participation in the new learning style become advocates for their children's education. Therefore, it is important to facilitate opportunities for parents to become familiar with the new expectations. Typically, students come home and parents ask them what homework they have and look for books as evidence of their children's commitment to doing the work. With a problem-based and self-directed approach, students decide what they need to do more often than doing more traditional, teacher-centered work. You may give them a long-term assignment of, say, 3 weeks, and with no intermediate work due, some will put project engagement off. With no textbook coming home, parents begin to wonder. Sending home the assignment and rubric so that parents can see them helps.

Adult/student/staff relationships are important points to consider when trying to capture credibility for classroom reform. Parents need to see reform as providing greater student opportunities for success. Change is threatening for many. Students are expected to begin to make sense of difficult material on their own. Students who are insecure about their intellectual abilities or teachers who fear comparisons to their classroom practices or work rate may present problems. Students also begin to recognize that with problem-based learning, they will not be able to shirk homework or juggle commitments and rely on their test-taking skills to bail them out. The structure of implementing reform is fragile. Within your educational community, who will facilitate or hinder your efforts? Soliciting and involving peers and their ideas may help. From a motivational speaker on reform, I have heard a metaphor for the classification of staff in school that characterizes them as trailblazers, pioneers, settlers, or saboteurs. Know with whom you are dealing. Within faculty comes support for your efforts as well as potential saboteurs who view your new curriculum and methodologies as threatening. Not everyone is going to love what you are doing. You need support, and there is no set formula for acquiring it; however, being proactive in planning for change may solve problems before they arise. Involve supportive people in the planning stages of your work.

One way to acquire the data necessary to credit your effort is to do a little research on your own. Research instruments, usually questionnaires or end-of-the-course student evaluations, can identify variables indigenous to secondary-level science classrooms that foster interest, motivation, and a positive attitude toward science and related fields. Teacher knowledge and management skills, the usefulness of course content, and instructional methods all contribute to the perception that students exhibit toward science education. These things can be measured and quantified over a few years. College class evaluations are a good source of questions or approaches to accessing attitudes. From past experience, teacher personality is the single most important factor affecting student attitudes. However, quantified preferences and attitudes toward instructional styles can be evidence of more interested, motivated, and engaged students in your learning and teaching model.

For additional help, in searching through educational journals, I have found that advocates for special education and gifted and talented education are a great source of ideas for assessment and evaluation of curricula.

A Word About Group, Cooperative, and Individual Study in the Science Classroom

Science, as practiced in the real world, can be very much a team game. Research and investigative progress are built on the work of others. Because of this, communication is a huge part of the scientific process. Reading journals, attending conferences, and general cooperation is a very important and huge part of doing science. In addition to this, science, for the most part, is self-regulated and self-assessed/ -evaluated by peer review. Bringing this to the classroom and building a sense of these processes into the curriculum can be important to the authenticity of the problem-based approach. However, at some time during the problem-based learning process, perhaps at the end of the first year, after students experience group problem-based learning and are more knowledgeable about strategies and reasoning, the group process may no longer offer educational advantages for all students. Also, self-motivated students do not always want to support group weaknesses, where less motivated students develop the habit of relying on others to avoid work. Sometimes, parents become frustrated with too much group work as well. Group work does not always offer individual opportunities to develop self-confidence in the total process. Certain individuals require less guidance and support from the teachers, and this allows teachers more time for those that need it. Curriculum planners need to be aware of and consider these factors when creating experiences and developing timelines. A balance may need to be facilitated between group and individual explorations. Individual responsibilities and accountabilities can be built into group work and curriculum planning. The complexity and rigor of the problem may drive the student/problem organization. Some problems simply require more students or a specialized team to solve or work through.

A well-planned experience defines not only group goals and objectives but also individual responsibilities, goals, and objectives. Both sets of expectations can be evaluated, and success can be measured. Sometimes, the size and complexity of the problem dictates the number of learning opportunities available. Adjustments can be made if resources become limited or tasks become unmanageable because of a variety of reasons.

Positive interactions between individuals can also occur informally and should be encouraged. Group meetings can be held to discuss individual approaches or resources, or how the individuals resolved the problems and what they learned. These all can be orchestrated by a problem-based approach.

The Curriculum Construction Kit for the Life Sciences: In Closing

In the spirit of problem- and thematic-based curricula, the information contained within *Doing Science: Innovative Curriculum for the Life Sciences* was put forth to act as a trigger for thought, evaluation, and choice within curricular design and development. It is certainly open for review and critique. It should not be held as evidence for the rightness or wrongness of any other classroom practice. *Doing Science: Innovative Curriculum for the Life Sciences* involves helping individuals fulfill their unique needs for the development of more meaningful curricula for their classrooms. For those looking for instructional resources, *Doing Science: Innovative Curriculum for the Life Sciences* provides a reference for critique and decisions involving their appropriateness. Greater choice and a rich variety of options are major objectives here.

Innovation and change can be messy, dangerous, time-consuming political business that has few short-term benefits for the proponent of change. Genuine change requires patience and, at times, resiliency. Teachers are given a variety of philosophies, guidelines, and frameworks in which to synthesize their own working curriculum. I have tried to provide the background, justification, and information that curriculum designers will need by revising, modifying, or reinforcing their current practice. We are developing future citizens above all else, and you, as a professional educator, have the power within your four walls to do it best.

Bibliography

Adams, C. M., & Callahan, C. M. W. (1995). The reliability and validity of a performance task for evaluating science process skills. *Gifted Child Quarterly, 39*(1), 14-21.

Barrows, H. S. (1985). *How to design a problem-based curriculum for the preclinical years.* New York: Springer.

Barrows, H. S., & Tamblyn, R. M. (1980). *Problem-based learning: An approach to medical education.* New York: Springer.

Corbett, D., & Wilson, B. (1995, June/July). Make a difference with, not for, students: A plea to researchers and reformers. *Educational Research, 24*(5), 12-17.

Delcourt, M. (1993, Winter). Creative productivity among secondary school students: Combining energy, interest, and imagination. *Gifted Child Quarterly, 37*(1), 23-31.

Denastes, S., & Wandersee, J. H. (1992, January). Biological literacy in a college biology classroom. *Bioscience, 42*(1), 63-65.

Eisner, E. W. (1985). *Educational imagination: On the design and evaluation of school programs.* New York: Macmillan.

Glasgow, N. A. (1996). *Taking the classroom into the community: A guidebook.* Thousand Oaks, CA: Corwin.

Good, M. L., & Lane, N. F. (1994, November 4). Producing the finest scientists and engineers for the 21st century. *Science, 266,* 741-743.

Goodman, J. (1995, Spring). Change without difference: School restructuring in historical perspective. *Harvard Education Review, 65*(1), 1-29.

Holden, C. (1994, December 9). National standards finally ready for public scrutiny. *Science, 266,* 1637.

Johnson, D. T., Boyce, L. N., & VanTassel-Baska, J. (1995, Winter). Science curriculum review: Evaluating materials for high ability learners. *Gifted Child Quarterly, 39,* N1.

Kaufman, A. (1985). *Implementing problem-based medical education.* New York: Springer.

Kingore, B. (1993, January-February). Gifted curriculum: The state of the art. *Gifted Child Today, 16*(84), N1.

Lewy, A. (1991).*National and school-based curriculum development.* Paris: United Nations Educational, Scientific and Cultural Organization.

Ost, D. H., & Yager, R. E. (1993, May). Biology, STS and the next steps in program design and curriculum development. *The American Biology Teacher, 55*(5), 282-287.

Schmidt, H. G., Lipkin, M., Jr., de Vries, M. W., & Greep, J. M. (Eds). (1989). *New directions for medical education.* New York: Springer-Verlag.

Seufert, W. (1993, July 10). Physician, teach thyself. *New Scientist, 139,* 41-42.

Shore, B. M., Koller, M., & Dover, A. (1994, Fall). More from the water jars: A reanalysis of problem-solving performance among gifted and nongifted children. *Gifted Child Quarterly, 38*(4), 179-183.

Speece, S. (1993, May). National science education standards: How you can make a difference. *The American Biology Teacher, 55*(5), 265-267.

Swanson, D. B., Norman, G. R., & Linn, R. L. (1995, June/July). Performance-based assessment: Lessons from the health professions. *Educational Research, 24*(5), 5-10.

Tobias, S. (1992). *Revitalizing undergraduate science.* Tucson, AZ: Research Corporation.

Uno, G. E., & Bybee, R. W. (1994, September). Understanding the Dimensions of Biological Literacy. *Bioscience, 44*(8), 553-557.

Index